PLC Controls with Structured Text (ST)

PREFACE

When I started as Assistant Professor (lecturer) at Dania Academy, Randers, Denmark, one of my first tasks was to find books for the students at the Automation Technology degree course.
It turned out to be particularly difficult to find relevant material related to programming in Structured Text (ST). Books with hundreds of pages would only contain a few pages about ST programming, and only at a very theoretical level.
My students enjoy learning ST programming from practical examples and methods. In January 2017, I began work to fill this gap by developing new teaching material under the working title of:

"Get Started with Structured Text"

Whilst using the material in my lectures, it has continuously been updated and extended. There has been high demand for the material among my students, and I have now turned it into a book for other interested readers to benefit from.

It is my hope that you will enjoy this book.

I would like to extend my gratitude to my students, fellow lecturers and colleagues for feedback and inspiration.

Comments, complaints, compliments and suggestions are welcome and appreciated.
Please, send them to TomMejerAntonsen@gmail.com

First edition issued June 2018

The 3rd edition has been updated and expanded with many of the suggestions and questions that readers and students have come up with, including the desire for many more illustrations and program examples.

Please enjoy!

Tom Mejer Antonsen

Randers, Denmark (June 2020)

Tom Mejer Antonsen

PLC Controls with Structured Text (ST)

IEC 61131-3 and best practice ST-programming

3. Edition, June 2020

Illustrations and graphics: Tom Mejer Antonsen

Translated by: Katrine Bay Madsen

The original version (Danish) 1. Edition issued March 2018

Publisher: Books on Demand GmbH, Copenhagen, Denmark
Printed: Books on Demand GmbH, Norderstedt, Germany

ISBN: 978-87-4302-636-5

Table of contents

1 Introduction

This book gives an introduction to the programming language Structured Text (ST) which is used in **P**rogrammable **L**ogic **C**ontrollers (PLC) and **P**rogrammable **A**utomation **C**ontrollers (PAC).

The book can be used for all PLC types and PLC brands following the open international standard IEC 61131 part 3: programming languages.

In a Siemens PLC, the programming language is called **S**tructured **C**ontrol **L**anguage (SCL). SCL may differ slightly from programming in ST.

The book systematically describes basic programming, including advice and practical examples based on the author's extensive industrial experience.

Explanations to the PLC programming code with an emphasis on writing stable, robust, readable, structured and clean code are included in the book. The aim of the book is to enable the reader to write PLC code, which does not require a specific PLC type and can be reused across multiple types of PLCs.

It is recommended to read the entire book to gain an overview of its content, and then use the book as a reference moving forward.

The book was developed for the full-time "Academy Profession (AP) Graduate in Automation Engineering" course and the part-time "AP Degree in Automation and Operation" course at the Dania Academy, Randers, Denmark.

The book is based on the IEC 61131-3 standard. PLC suppliers and manufactures interpret the standard in different ways and not all follow the standard consistently. This means that some of the program examples in this book may not work properly in the PLC type you are using.

Unfortunately the author is not available for support in connection with programming code within this book.

1.1   Background for ST

ST is a high-level programming language similar to Pascal Programming. Pascal Programming was widely used from 1980 to approx. 2000 – a period in which many companies started developing software for PC running on DOS (Disk Operating System), and later software running on Windows.

ST was developed and published by the International Electrotechnical Commission (IEC) in IEC 61131-3 International Standard in 1993. The standard consists of five PLC programming languages of which the Ladder Diagram (LD) language is the most well-known and commonly used. In addition to ST and LD, the other PLC programming languages include **F**unction **B**lock **D**iagram (FBD), **I**nstruction **L**ist (IL) samt **S**equential **F**unction **C**hart (SFC).

Since about 2010 the usage of ST programming for PLCs has become widespread across Denmark, and many companies are now purely delivering PLCs programmed in ST. This means that the demand for ST programming capabilities across the industry has increased. This book is part of educating a workforce to fill this demand.

1.2   Prerequisites for learning ST programming

It is not necessary to know how to program in LD or the other PLC programming languages when learning ST programming. However, a certain level of knowledge of mathematics, mechanics, electronics, automation solutions and basic PLC is required to be able to learn ST programming.

Students with knowledge of a high-level programming language (e.g. VB, C++, C#, Python) will be able to learn ST relatively easy, due to the similarities in coding structures. The program execution inside a PLC is different compared to a traditional program or App running on a PC or a smartphone.

Like other text programming languages, the student can expect to be proficient in the language within three to five years.

1.3   Foundation of knowledge

The author has 25 years' industrial experience with the specification, development and delivery of complex control systems and supervision systems. Of the 25 years, the author has 7 years' experience with Pascal Programing and 12 years within automation solutions and systems involving PLC. This experience includes employment in four international companies and delivery of more than a thousand control system solutions in 20 different countries. This experience provides an important base for the content of this book

In recent years the author has been teaching PLC Systems at degree level in Denmark. The students have from 0 to 20 years of practical and/or vocational experience within PLC, automation and technological services. The internet, the standard DS/EN 61131-3 and a series of books on PLC programming have been utilized as inspiration for this book.

The material for this book was developed with feedback from lecturers and students attending the "Academy Profession (AP) Graduate in Automation Engineering" course and "AP Degree in Automation and Operation" course at the Dania Academy, Randers, Denmark. The content has been updated to answer the questions which the students typically ask during the course.

The author has a Bachelor of Science in Electrical Engineering (B.Sc.E.E.) from Aarhus University School of Engineering, Denmark.

1.4 Advantages of ST programming

ST is a flexible and universal programming language. As ST programming code is based on text and not graphics like LD, the code can easily be copied between different PLC types, and even be sent by e-mail.

The ST programming code is similar to text sentences, and work is executed in the same way as a word processor program such as Microsoft Word which makes it easier to work on. Consequently, the same working methods are applied when using a word processor program or a text editor.

Because of its very structured nature, ST is ideal for tasks based on complex mathematics, code reuse or decision-making (e.g. automatic energy optimization, algorithms, data collection and regulation in process plants).

Having the experience with PLC Programming, transitioning to other programming languages within PLCs and automation will be easier including robotics or Visual Basic programming.

Within recent years an increasing number of companies have switched to ST programming. This is due to a number of advantages provided within the ST programming language compared to the four other PLC programming languages (LD, SFC, FBD and IL).

The advantages of the ST programming language are:

- ST Programming code can relatively easily be copied between different PLC types and brands[1].

- It is the most convenient PLC language for mathematical calculations, formulas and algorithms[2], and for managing large amounts of data (Big Data).

- PLC solutions are more in demand today than 20 years ago[3].

- Many widespread programming languages (C++, C#, VB, PASCAL) share similarities with the ST program structure.

- Other PLC languages (LD, SFC, and FBD) require parts of the program to be written in ST anyway.

- Documentation of the ST PLC programming code requires less space during documentation, description and printing compared to other PLC programming languages.

- It is the easiest PLC language to version control via comments in the program code or via GIT or Subversion[4].

The PLC programming language Instruction List (IL) which is applied for complex PLC Controls is expected to be outdated within year 2020/2021 (cf. IEC 61131-3 section 7.2.1). It is expected that ST will replace these solutions.

---

[1] This is possible by using copy-paste and minor corrections. For example, a Siemens PLC uses the sign # before naming local variables, and Allen Bradley PLC uses a different syntax to make function 'calls'.

[2] Mathematical calculations are similar to mathematical formulas. Chapter 8.1 page 47.

[3] There is more focus on energy optimization, automatic operation and data collection today. These are all solutions which requires more complex PLC coding than an ordinary 'relay/circuit breaker' with start/stop functions.

[4] The tools GIT and Subversion are practical tools which allows the user to track (follow) corrections and extensions in the PLC Code. This makes it possible to commit current changes and fetch earlier versions of the PLC Code.

1.5   Disadvantages of ST programming

A big disadvantage is the fact that many technicians and electricians are only capable of programming in LD. It is difficult for them to understand ST program code because it is written in text and is not graphical like the LD program code[1].

As a certain level of experience in structuring a program is required, programming in ST can easily be confusing.

Inexperienced people may have difficulties in fault-finding (debug) in an ST program.

Small (Micro) PLCs do normally not allow ST Programming.

It is not normally possible to apply ST Programming in a safety PLC[2].

Reaching expert level in ST programming often takes three to five years upon completion of a formal course or education.

2   How the PLC executes PLC code

It is important to know how the PLC executes the program code when writing the program. A PLC executes the program in real-time, which means that the program modules must be executed within a short time-frame. The program modules are executed at a fixed time interval (the PLC scan time) e.g. 50 [ms]. Some of the fastest PLCs may have a scan time of 1 [µs].

Program modules can have different scan times e.g. 500 [ms] or 1000 [ms]. Some sensor input values do not change quickly (e.g. a temperature sensor). Which means that it is not necessary to have fast scan time for all program modules. A large program with many calculations takes a longer time to execute. This means that different programs will require different lengths of scan time. See also chapter 10.1, page 78.

---

[1] To help learners who are already proficient (good) in the LD programming language, and would like to start programming in ST, Chapter 14 page 178) will be a good starting point. The chapter provides examples of different programs written in LD and the equivalent using the ST.

[2] A safety PLC is a separate PLC or a dedicated area in an ordinary PLC used to ensure the stop of the machine or plant when the emergency stop device is activated.

The basic mode of operation for a PLC:

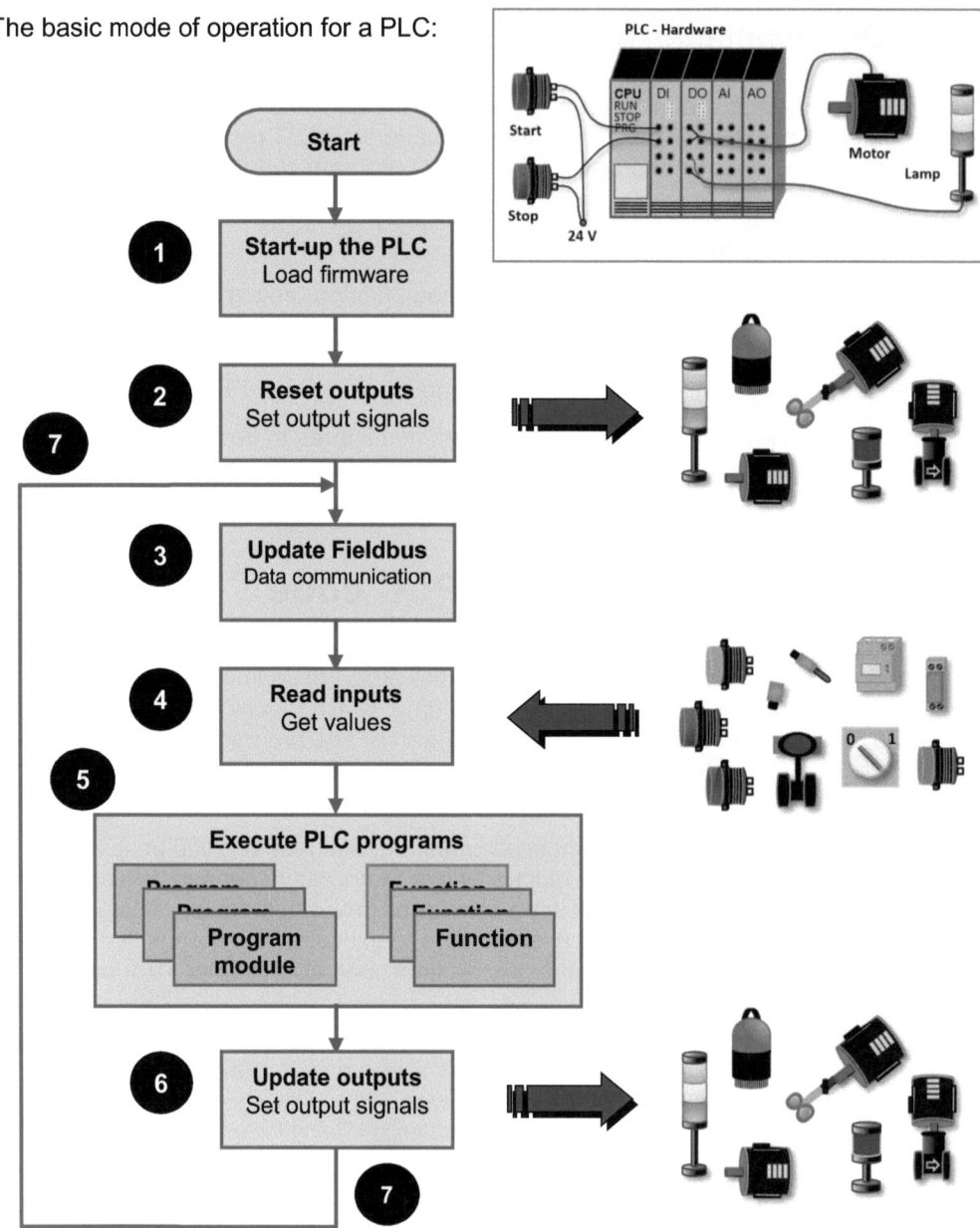

The flow diagram shows the following points:

1 When power is turned on the PLC will start (boot up) and load the operating system, called firmware in a PLC system. This will ensure that the PLC program knows the connected hardware (HW).

2 After startup, all output modules are set to its initialization values. It is important that all outputs have the correct startup values so there are no unintended actions before the PLC program has started.

3 A data communication link has now been created via a network (fieldbus). Variables are received and sent out to other units (e.g. control panels, other control systems or instruments). There are many types of fieldbus network systems. Some of the most used include Profibus, Profinet and Ethernet/IP. However, most fieldbuses are built with similar functionality, and they all work in similar ways.

4 Values from all sensors, switches, instruments and components on the machine or plant are now received from the connected input modules.

5 Dependent on the scan time all PLC programs will be executed once. Programs are split up as follows:

> Program modules. See chapter 10.1, page 78
> Functions. See chapter 10.2, page 80
> Functions (FC) and Functions blocks (FB). Chapter10.3, page 82

Programs must be split up in order to create a good program structure.

6 Values are written to all output modules. Values could include new settings to motors/engines, valves, lamps and instruments.

7 Step 3 to 6 will be repeated. This is one program scan.

The execution of the program only stops if:

> - The PLC is set to STOP operation mode
> - If a run time error in the program occurs
> - The PLC is powered off or loses power unintentionally

3 Comments in the programming code

Comments are a very important part of programming. Comments in the programming code assist you and your colleague when later adding to the code.

Use comments to explain what a specific PLC code performs, so you can remember it later yourself. In many cases, the PLC code can be self-explanatory, therefore it is best only to make comments when the code is complex.

There are two types of comments in ST:

Line Comment:

```
// Line comment. Forward-slash is written in front of EVERY line.
// Here you can write comments

//You can write comments before a code section starts
IF S1 THEN
    K1:= TRUE;  //Or write comments after the code at the same line
END_IF;

//Do not execute the code below
//K2:= B2
```

Line comments can also be used to comment out PLC code so that it will not be executed. The code is lost if it is deleted, therefore place // in the beginning of the line instead of deleting the code. By doing this, the code is still available, but not executed

Line comments can only be placed on the same line in front of or after code.

Block Comment:

```
(* Block comment is initiated by a start parenthesis and a star. It is finalized by
a star and end parentheses. They are used for making more lines of PLC code
inactive i.e. create multiplecomment lines *)
```

Comments placed between (* and *) are called block comments and are used in order to remove/sort out more lines of code or to write comments filling up more lines.

Comments are essential at the start of any program module or function. This ensures that other programmers can understand the intended functionality of the program quickly and easily, without having to trawl through the entire code.

It is best practice to maintain a version log at the top of the program module which should include any changes made to code and the author of these:

```
///////////////////////////////////////////////////////////////////////////////
/// OP002 Parking house
///////////////////////////////////////////////////////////////////////////////
// Action for each connected sensor
//
//************************************************
// Version 1.0, Created. Date 06.05.2020 TMA
// Version 1.1, TempVar3 changed 10.05.2020 TMA
// Version 1.2, Button S1 added 1.05.2020 TMA

IF S1 THEN    //First line of PLC code
   K1:= TRUE;
END_IF;

//SetLamps();  //Do not run the SetLamps program module
```

A few PLC types cannot handle the special localized language characters such as æøå/ÆØÅ in the comment lines. As localized characters are not accepted by the PLC programming tool, It is therefore recommended to use the English alphabet in both the comment lines and the programming code. Due to this many companies choose to write their PLC code in English.

IMPORTANT!   Do remember to correct the comments and version log
                   if anything is later changed in the PLC code.

TIPS: Before starting the actual programming phase, you can use comments
              to gain a better understanding of the intended functionality of the pro-
              gram. This could help you to achieve more structure within the code,
              and help readers of the code understand it more easily.

4   Data types

Just like other programming languages, the IEC 61131-3 programming standard provides many different data types which include both elementary and complex ones. A data type defines how much memory capacity is needed by a variable value and by that, the largest and smallest value in the variable.

4.1   Elementary data types (INT, REAL, BOOL)

The following (examples) simple data types are standard in any PLC controller:

Data type	Bits	Numeral system	Note	Lowest and highest value	Eksempel
BOOL (Bit)	1	Boolean (Binary)		**FALSE/TRUE** or 0 to 1	**TRUE**
BYTE	8	HEX (Hexadecimal)		16#0 to 16#FF	16#10
WORD	16	Binary		2#0 to 2#1111111111111111	2#0001000000000000
UINT	16	HEX (Hexadecimal)		16#0 to 16#FFFF	16#1000
		BCD (Binary-Coded Decimal)		C#0 to C#999	C#998
		Unsigned Integer (positive numbers)		0 to 65535	564
DWORD (Double word)	32	Binary		2#0 to 2#11111111111111111111111111111111	2#10000001000110001011101101111111
		HEX (Hexadecimal)		16#00000000 to 16#FFFFFFFF	16#00A21234
		Unsigned Double word (integer)		0 to 4294967295 (4.29 billion)	435
INT (Integer)	16	Signed integer		-32768 to 32767	101
DINT (Double integer)	32	Signed double integer		-2147483648 to 2147483647 (2.1 billion)	107

Data type	Bits	Numeral system	Note	Lowest and highest value	Example
REAL (Floating-point number)	32	IEEE 754 Floating-point number (Decimal tal)	1	Lowest value: +/-3.402823E+38 Highest value: +/-1.175495E-38	1.234567e+13
LREAL (Long Real)	64	Dobbelt Float (Decimal tal) IEEE 754		Lowest:-1.7976931348623E308 Highest: 1.79769313486232E308	3432.54
TIME (IEC time) **LTime**	32 64	IEC time Step in 1 [ms] or Step in 1 [ns]	4	T#1ns to T#24d20h31m23s	TIME#10s T#10d14h11m23s T#5s12ms23us300ns
DATE (IEC date)	16	IEC day, step 1 day		D#1990-1-1 to D#2168-12-31	D#1996-3-15 DATE#1996-3-15
TIME _OF_DAY (Time)	32	Time in a step of 1 [ms]	4	TOD#0:0:0.0 to TOD#23:59:59.999	TOD#1:10:3.3 TIME_OF_DAY#1:10:3.3
CHAR **WCHAR**	8 16	ASCII characters (letter or sign)	2	'A', 'B' etc.	'E'
STRING		Text	3	Up to 255 characters	"This is a text"

All variables must have a data type. If a variable is given a value outside the minimum and maximum value range of the data type, a run time error may occur and consequently the PLC may stop the program execution. This may again lead to strange behavior when executing the program (the program may seem unstable).

A few PLC types provide more data types than the ones listed above. In general, it is recommended use only a few data types so that the PLC code can be copied in an easier way to other PLC types. Some special data types such as **S7TIME**, **LWORD** and **ULINT** cannot be used by all PLC types. This means that copying PLC code with special datatypes, or upgrading to a larger PLC, may take a lot of work and risk introducing errors to the code.

The three most used data types are **BOOL**, **INT** and **REAL**. The reason why **INT** is used more often than **WORD** is that **INT** provides the same amount of data as the bit-size in a PLC making it a fast data type. On the other hand, if **REAL** is used, the PLC will auto-generate underlying machine code as the PLC can only work with integers.

The disadvantage of working with **INT** is when exchanging values between computers where e.g. one computer is a PLC running on a 16-bit operating system, and the second is running on a 64-bit operating system. The second computer could also be a small 8-bit computer (an embedded computer), used inside a sensor, a measuring instrument or equipment for analyzing process values. Read more in chapter 8.5, page 51.

Data type table notes	
1)	A **REAL** integer contains at most 7 influential digits. This means that if a variable is allocated the value of 1234.56789, the variable is not able to contain all digits. The value will consequently be changed to the value of 1234.567 (7 digits). Some PLC types use 8 digits: 1234.5678.
	In some PLC types these data types are named **FLOAT**.
	Because computers may handle a **REAL/FLOAT** differently, some challenges can occur when communicating between several computers. In order to handle this, a **REAL** can be changed to an **INT** or **DINT** variable by multiplying by 100, and when data is received in another computer the variable has to be divided by 100. Using this method, a decimal number including 2 digits can be transferred without any problems. See more in Chapter 8.5, page 51.
2)	ASCII characters are typically used when texts are needed to be written on e.g. user interfaces, data logging to files, communication between instruments, data from a keyboard or other PLCs. Due to the fact that a PLC operates with integers only, letters and signs each have a number in an ASCII table.
	The data type **CHAR** has 8 bits (may contain 255 different characters). A **CHAR** data type may typically be used for 1 to 5 different languages (countries). **WCHAR** has 16 bits and is applied for Unicode (ISO 10646, Universal Coded Character Set). Unicode is used for international PLC solutions.
	WCHAR is typically used when the same PLC-code is applied in several countries with different languages in the user interface.
3)	A **STRING** consists of an **ARRAY** of **CHAR** and is normally set to 255 characters (**CHARS**)
	See above-mentioned note 2).
	Furthermore, see chapter 11, page 94.
	WSTRING is applied for Unicode (ISO 10646, Universal Coded Character Set) and consists of an **ARRAY** of **WCHAR**.
	Note: Some PLC types provide a maximum of 80 characters in a **STRING**, if the **ARRAY** is not limited e.g. 10. It is good practice in programming to limit ARRAY so that unnecessary memory is not used.

4)   **TIME/DATE** is calculated internally in a PLC as an integer, which counts time from 1.1. 1970 at 00.00 and can therefore only be converted to an integer. (See the documentation from the individual PLC-type)

A PLC gets its current time from an in-built electronic component in the PLC hardware. However, its time indication is not very accurate. An accurate time indication must be fetched from an atomic clock, which allows a PLC to be fully automatic straight away if connected to the internet. A PLC can then get its current time from an ordinary PC, e.g. once a day. It is important that all PLCs on the network show the same time so that alarms and stamping with date and hour of logged data indicate the same time (e.g. event log – logging of changes made by the user in the PLC control).

When a variable is assigned a value (set to a value), the value is normally (by default) a decimal number. If the value is a binary number, **2#** must be written in front of the number, and if it is a HEX number, **16#** must be written in front of the number. E.g. **2#**101 = 5 or **16#**FF = 255.

When deciding on what data type to use for a variable, it is important to know its maximum value capacity. Normally an **INT** data type is used for counters. If **INT** is used as TACHO HOURS on a motor, the maximum value of an INT can be a problem. TACHO HOURS is a counter showing the total number of hours a motor has run, and is used to indicate when the service interval has ended, and motor service is required. If for example the motor runs 20 hours a day, and it has an expected life time of 10 years, the total counter value will be reached as follows:

Hours within 24 hours*days per year*year = 20*365*10 = 73,000 (hours)

The problem is that the variable cannot be contained in the data type **INT**, as **INT** has a maximum value of 32767. A double integer **DINT** must be used instead, or even better a **DWORD** data type as it is able to contain an even larger number.

DWORD may contain an integer value from 0 to 4.29 billion.

If an **INT** is used anyway, the variable will show: 7466 as the **INT** has two 'overflows. An 'overflow' takes place every time the integer is higher than 32767 and at an 'overflow', the variable is reset to -32768 (which is the lowest value for **INT**).

4.2   User defined data types

It is possible to define more advanced and complex data types to save time when programming, and to obtain a better program structure. The data types are named user defined or derived data types and are declared within **TYPE** and **END_TYPE**.

There are three user defined data types: **ENUM** (Enumerated data type), which is a list of constant numbers. **STRUCT** (Structured data type) which group different variables in a structure. **ARRAY** contains a series of variables having same data type.

NOTICE
If an absolute beginner starts programming in a PLC, it is important to know that the derived data types are not necessary to use to make PLC programs work. Only start using derived data types when greater experience in PLC programming is gained.

The different user defined data types are explained in the following chapters.

4.3   Enumerated data type, ENUM

The enumerated data type ENUM contains a list of unique names. Names are listed in parentheses, and must be meaningful with regard to their purpose.
The declaring begins with **TYPE** and ends with **END_TYPE**.

Example:

```
TYPE LightTYPE :
      (RED, YELLOW, GREEN);
END_TYPE
```

The data type **LightTYPE** in the example above can either be RED, YELLOW or GREEN. **LightTYPE** could be used to control a traffic light, an operator signal lamp, a light tower (see picture) on a machine or as a status on a valve.

LightTYPE will always take one of the defined types: RED, YELLOW or GREEN.

An ENUM must be allocated a default value. A default value is required to ensure the right value during start-up (initialization). In the example below **LightTYPE** is initialized with the default value RED when the PLC is powered up:

```
TYPE LightTYPE :
     (RED, YELLOW, GREEN):= RED;
END_TYPE
```

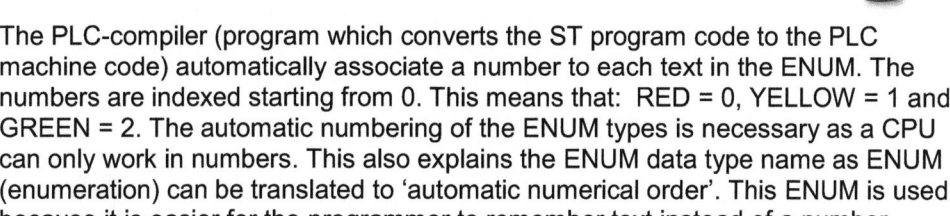

The PLC-compiler (program which converts the ST program code to the PLC machine code) automatically associate a number to each text in the ENUM. The numbers are indexed starting from 0. This means that: RED = 0, YELLOW = 1 and GREEN = 2. The automatic numbering of the ENUM types is necessary as a CPU can only work in numbers. This also explains the ENUM data type name as ENUM (enumeration) can be translated to 'automatic numerical order'. This ENUM is used because it is easier for the programmer to remember text instead of a number.

It is possible to define a fixed value for each name instead of automatic:

```
TYPE LightTYPE :
     (RED:= 10, YELLOW:= 20, GREEN:= 30) := RED;
END_TYPE
```

The disadvantage of using ENUM is that all numbers are positioned in a continuous order (indexed). If new names are added in the middle of the sequence, the index is disrupted which will cause issues when ENUM variables are exchanged between more PLCs or computers, as all devices must be updated with the new PLC code at the same time.

Examples of use: Below are two variables, **MotorLamp** and **Lamp**, both having the data type **LightTYPE**:

```
Lamp:= MotorLamp;              //Here is Lamp set to red
MotorLamp:= LightTYPE.GREEN;   //Set MotorLamp to green
Lamp:= MotorLamp;              //Here is Lamp set to green
```

ENUM creates a better structure, but ENUM is not possible in all PLC types.

The alternative to ENUM is to use independent constants. See chapter 6.2, page 38.

4.4   Structured data type, STRUCT

A structured data type, **STRUCT**, is a composite data type used to group more datatypes in a class/object. The structured data type is declared by using the key words **TYPE**, **STRUCT** and **END_STRUCT**.
Each variable in a **STRUCT** needs to have a name followed by a colon, and then the data type. Note that the declaring is ended by a semicolon.

Below a **STRUCT** is shown called **Motor**, containing four variables which are all related to a motor. **Speed** (Motor speed), **Temperature** (measurement inside the motor), **Voltage** (Power supply for the motor) and **AlarmStatus:**

```
TYPE Motor :                        //Example 1 STRUCT
  STRUCT
    Speed         : INT;    //Actual speed of the motor [RPM]
    Temperature   : REAL;   //Temperature inside the motor [C]
    Voltage       : REAL;   //The voltage of the motor [V]
    AlarmStatus   : BOOL;   //Alarm if TRUE else FALSE
  END_STRUCT;
END_TYPE
```

Motor

Note that comments are written after each variable which accurately describe the functionality of the variable to the reader of the PLC program. Furthermore, a unit is written in square brackets because the unit of different variables is often not known. For example, the speed of a motor could be measured in RPM (revolutions per minute), the frequency in Hz (Hertz) or in percentages (0 to 100%).

When the variable is declared, comment lines are also used to describe the behavior of the variable, as this is not always obvious or logical; e.g. the **AlarmStatus** where it is not clear whether the alarm goes off when the variable is **TRUE** or **FALSE**.

As mentioned in chapter 6.1, page 36, the unit can be a part of the variable name.

Some PLC types do not use text, as in the example above, when declaring a **STRUCT**; instead they are declared (written) in a list and therefore the key words: **TYPE**, **STRUCT**, **END_STRUCT** or **END_TYPE** will not appear to the reader.

A structured data type may contain one or more other derived data types. This can be seen in the example below:

```
TYPE Valve :              //Example 2 STRUCT
   STRUCT
   DisplayColor  : LightTYPE;   //User defined TYPE
   ValveState    : BOOL;        //Can be TRUE (open)
                                //or FALSE (closed)
   Pressure      : REAL;        //Pressure in [Bar]
   END_STRUCT;
END_TYPE
```

Ventil

In example 2 above the data type **Valve** consists of tree variables:

DisplayColor, **ValveState** (status of the valve: open or closed) and **Pressure**. The variables **Pressure** and **ValveState** use the standard data types **REAL** and **BOOL**, while the variable **DisplayColor** uses the data type **LightTYPE**, which is defined in chapter 4.3, page 18.

Example of a portable tank containing chemicals (IBC tank):

```
TYPE TankType :           //Example 3 STRUCT
   STRUCT
   Liters   : REAL := 1000;   //Default tank size
   LevelSensor   : REAL;      //Sensor at bottom
   LevelSwitch   : BOOL;      //Float switch at bottom
   END_STRUCT;
END_TYPE
```

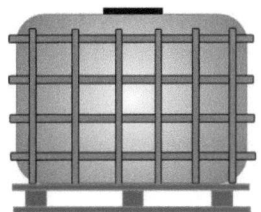

Many variables in a PLC program can easily become confusing. Variables belonging to the same component (object), the same domain, or the same mode of operation may advantageously be grouped in a **STRUCT**. Grouping variables makes it easier and quicker to set up and maintain many identical components. This method of programming is called Object Oriented Programming (OOP), and is often used when writing computer programs.

If a variable with the data type **STRUCT** is to be transferred to a function the variable scope must be set to **VAR_IN_OUT** within the function. See chapter 5, page 28.

4.5   Collection of values with same data type, ARRAY

An **ARRAY** is a structure, which can store a collection of values with *the same data type*. The values are located side by side in memory which means that it is simple to work on. An **ARRAY** always has a fixed length which cannot be changed during the execution of the program. An **ARRAY** can be set up and indexed by several dimensions.

You can write **ARRAY** programming code quickly, and it provides a good programming structure. The challenge is getting the values in and out of the **ARRAY**.

An **ARRAY** is also called a multi elementary data type.

Below example show an **ARRAY**, **SpeedArray**, which contains 6 positions of the data type **INT**. To declare the 6 positions, use **ARRAY** followed by square brackets including start end position number, separated by two dots as shown below:

```
VAR SpeedArray :
    ARRAY [1 .. 6] OF INT;
END_VAR
```

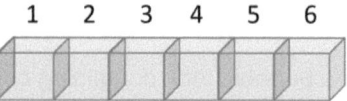

The first value in the array is located in position no. 1 and the last in position no. 6. The name for the **ARRAY** in this example is **Speed,** which is added to the text **Array,** so that any person working on the PLC code will easily know that an **ARRAY** is used.

SpeedArray is a one-dimensional **ARRAY** and can be used where a collection of many values with same data type is positioned in one long row. Examples include:

Calculation of the average value (chapter 10.4.2, page 74).
Handling of a queue (chapter 13.1, page 126).
FIFO - First In First Out (chapter 13.2, page 129).
Collection of data and sorting (chapter 9.4.5, page 77)

An **ARRAY** can be used with all data types, including **STRING**, **STRUCT** or functions.

Examples of the use of **ARRAY** can be found on pages 72, 74 or 126.

A two-dimensional **ARRAY** can be used on e.g. a parking lot (car park), stock rack, a graph, a bar chart or a pivot table and can be set up as follows:

```
VAR Racking
    ARRAY [1 .. 5, 1 .. 3] OF INT;
END_VAR
```

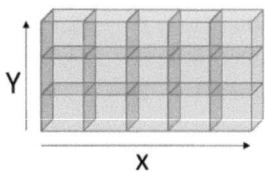

A three-dimensional **ARRAY** is defined as follows:

```
VAR PackOnPallet
    ARRAY [1 .. 5, 1 .. 4, 1 .. 3] OF REAL;
END_VAR
```

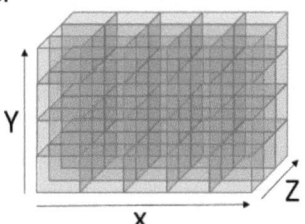

Used e.g. for packages on a pallet (palletizing) or positions in a warehouse.

If you look at a three-dimensional **ARRAY** as an X, Y and Z system of coordinates, the values from the above-mentioned example can be grouped as follows:

X = 1 til 5,   **Y** = 1 til 4,   **Z** = 1 til 3.

The total amount of positions in **PackOnPallet ARRAY** is: 5*4*3 = 60 pieces. So this **ARRAY** contains 60 positions (elements).

An **ARRAY** can be defined with an index starting point of 0. The **ARRAY** below contains 4 positions as position 0 (zero) and position 3 are included when counting the number of positions in the array. It results in a more stable program when arrays start from 0, because the array index pointer remain uninitialized (not given an start value):

```
VAR MyArray1D
    ARRAY [0 .. 3] OF INT;
END_VAR
```

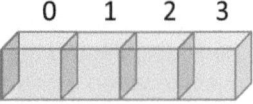

Insert a single value in an ARRAY
In the below example the value 5 is inserted at position 4 in the one-dimensional **ARRAY SpeedArray**:

```
SpeedArray [4] := 5;
```

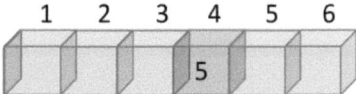

Values can be inserted in the three-dimensional **ARRAY PackOnPallet** as follows:

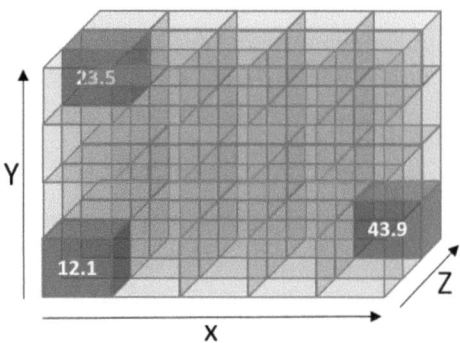

```
PackOnPallet [1, 1, 1] := 12.1;
PackOnPallet [5, 1, 3] := 43.9;
PackOnPallet [1, 4, 2] := 23.5;
```

For inserting multiple values in a 3D **ARRAY**, see chapter 9.4.2 page 73.

Get a value from an array

In this example shows how to get a value from a one-dimensional **ARRAY**. The value is located at position 2 in the array **MyArray1D** and copied to the variable **Var1**:

```
Var1 := MyArray1D [2];
//Contents of Var1 is 12
```

Get a value from a three-dimensional **ARRAY** is carried out as follows:
A value is transferred (copied) to the variable **Var3** with the value of 43.9:

```
Var3 := PackOnPallet [5, 1, 3];
//Contents of Var3 is 43.9
```

IMPORTANT: You must not assign (copy) a value to positions outside of the **ARRAY**. If assigning a value to for example, position no. 10 in an **ARRAY** containing only 6 positions, the PLC can stop the program execution (Run Time Error). This is a common error/mistake when using arrays for your code. To avoid assigning values to positions outside the **ARRAY** use an **IF** statement as shown below:

```
Index:= 4;  //Insert 5 at position 4
IF Index > 0 AND Index <= 6 THEN
    SpeedArray [Index] := 5;
END_IF;
```

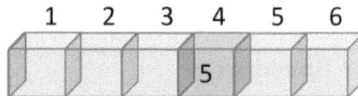

In the previous example, the low bound and upper bound of the array are used directly. This can be a disadvantage when the array is used inside a function or the lower and upper bounds have to be changed. Therefore, the built-in standard functions **LOWER_BOUND** and **UPPER_BOUND** can be used.
The two functions return the bound of the array and can be used as shown below:

```
Index:= 4;  //Insert 5 at position 4

IF Index >= LOWER_BOUND (SpeedArray, 1)
   AND Index <= UPPER_BOUND(SpeedArray, 1) THEN
   SpeedArray[Index]:= 5;
END_IF;
```

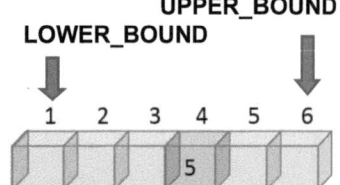

Direct and indirect addressing

For direct addressing in an array, numbers are used to retrieve the content at a particular location in an array and for indirect addressing, a variable is used:

```
//Indirect addressing is using a variable
Index:= 4;  //Index is an INT or a WORD data type
SpeedArray[Index]:= 5;

//Direct addressing when a number is used directly
SpeedArray[4]:= 5;
```

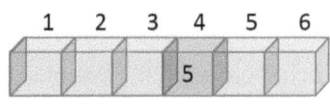

Data collection

An array is perfect for collecting data (data log) in a PLC controller.
This code collects the number of items produced from a machine:

The array is configured as shown:

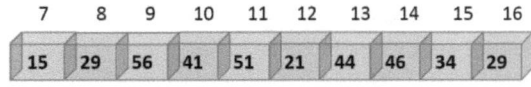

```
VAR
   DataCollect ARRAY [7 .. 16] OF WORD;
END_VAR
```

The array saves data from a range from 7 to 16 which corresponds to the period during which the machine produces items. From 7 o'clock to 8 o'clock the machine produces 15 items, between 8 o'clock and 9 o'clock the machine produces 29 items, etc.

By collecting the number of items produced, it is easy to find out the periods in which the machine has produced the most and the fewest items. The data collected confirms how efficient the production is during different time periods.

An element in an array can only save one value. This example shows how an array can contain multiple values, and how constant values are saved in an array.

The example is a warehouse rack, controlled by a robot:

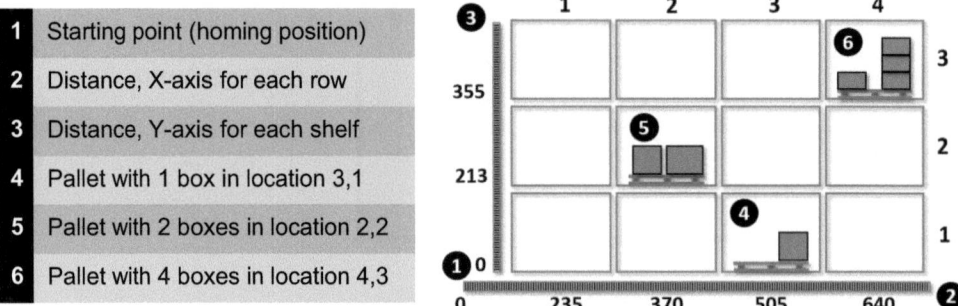

1	Starting point (homing position)
2	Distance, X-axis for each row
3	Distance, Y-axis for each shelf
4	Pallet with 1 box in location 3,1
5	Pallet with 2 boxes in location 2,2
6	Pallet with 4 boxes in location 4,3

Numbers at X-axis (2) and Y-axis (3) are predefined numbers to inform the robot how far to go (distance) to place an item on a certain row and shelf in the warehouse rack. Each axis uses an encoder to determine where a location is. An encoder is a sensor that sends a pulse for distance traveled, and when the correct number of pulses is received, the robot stops at the correct location.

Each location in the warehouse rack contains several values and therefore a **STRUCT** is created as shown to the right:

To add more values, add additional code lines under the **Weight** variable.

```
TYPE LocationTYPE
  STRUCT
  NoOfBox : WORD;
  Weight   : REAL;
  END_STRUCT
END_TYPE
```

To handle the warehouse rack, the following variables are created:

```
VAR CONSTANT
  StockSizeX: INT := 4; //Size x of the Stock
  StockSizeY: INT := 3; //Size y of the Stock
  StockEncodeX: ARRAY[1.. StockSizeX] OF INT := [235, 370, 505, 640];  //Encoder X values
  StockEncodeY: ARRAY[1.. StockSizeY] OF INT := [0, 213, 355]; //Encoder Y values
END_VAR
VAR
  Stock: ARRAY[1.. StockSizeX, 1.. StockSizeY] OF LocationTYPE; //Location is a STRUCT
END_VAR
```

Inserting values for the pallet located at (6) is done like this:

```
Stock[4, 3].NoOfBox := 4; //Located at StockEncodeX[4] and StockEncodeY[3]
Stock[4, 3].Weight := 1210.25; //Set weight
```

This example is based on a depalletizer:

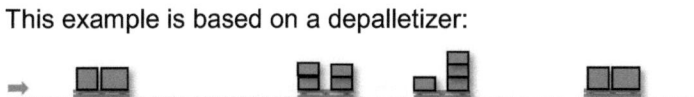

M4   M5   M6

M1

A conveyor belt can have different states and these are grouped in an **ENUM**:

```
TYPE ConveyorState :
  ( NONE, STOP,
    RUN_CW,          // Run clock wise
    RUN_CCW,         // Run counter clock wise
    ALARM) := STOP;  // Default set to stop mode
END_TYPE
```

By default the state is set to STOP to avoid the conveyor belt running unintentionally when the PLC is powered up.

The variables for a conveyor belt are grouped into a **STRUCT**:

```
TYPE ConveyorTYPE :
  STRUCT
    State      : ConveyorState;  //State/mode
    Speed_m_s  : REAL;    //Conveyor speed in [m/s]
    Size       : INT;     //Conveyor size, 40 or 60
  END_STRUCT
END_TYPE
```

Declaration of the conveyor belt variables and code examples are shown below:

```
VAR
  M1         : ConveyorTYPE;                    //Single conveyor
  ConveyALL : ARRAY [4..6] OF ConveyorTYPE; //All conveyor
END_VAR
```

```
//Start M1 single conveyor
M1.State:= ConveyorState.RUN_CW;
//Set size of conveyor 5
ConveyALL[5].Size:= 40;
//Start conveyor 5
ConveyALL[5].State:= ConveyorState.RUN_CW;
//Copy all variables from conveyor 5 to conveyer 6.
ConveyALL[6]:= ConveyALL[5]; // Conveyor 6 is a copy of conveyor 5
```

5   Variable scope

Variables are key elements in programming. All variables must have a data type.
When a variable is created (declared) it must be configured to use a variable scope.
The variable scope sets the value's behavior in the memory.
A table of the most common variable scopes in the PLC program is shown below:

Scope	Description
VAR	All local variables are declared between the keywords VAR and **END_VAR**. The local variables cannot be manipulated from outside the program module or the function. **NB**: In some PLC types **VAR** is replaced by *"Static"*
VAR_GLOBAL	Global variable scope. Variables in this scope can be accessed (called) from all program modules, functions, Fieldbus (networks) and HMIs (user interfaces). The use of global variables should be kept to a minimum as it makes the PLC code more complex and harder to find errors.
VAR_INPUT	Used by functions for variable input into a function. See more in chapter
VAR_OUTPUT	Used by functions to *return* variable values after the value has been worked on/changed by the function. See more in chapter 10.2 page 80.
VAR_IN_OUT	Input and output variable scope for functions. An address (a link) of the variable which is *transferred* to the function. Changes are made directly to the variable and *not a copy* of the variable as is the case when using **VAR_INPUT**. Used when a function has to work with a **STRUCT** or **ARRAY**. This scope must be used carefully as the function changes variables located outside the function. See more in chapter 10.2 page 80.
VAR_EXTERNAL	If a program module uses this scope on a variable, the program module will be able to use the global variable of the same name. Must be used with caution.
VAR_TEMP	A temporary variable scope in the function which means that the contents of the variable disappears when the function is finished. **NB**: In some PLC types **VAR_TEMP** is replaced by *"Local Temp"*.

Scope	Description
AT	Defines a memory location (mapping address) for a variable.
	An example of this could be an **I/O** address (the address on a PLC **I**nput or **O**utput). The input could be named %IX 1.0, where %I indicate that it is an input. The output name could be %QX 0.0, where %Q indicate that it is an output.
	Q is used as a letter for output (O is not used as it can be confused with zero/nil).
	See the example in chapter 5.1, page 28.
	If a memory location is not definded, the PLC will automatically allocate the next free internal address in the memory.
CONSTANT	Variables cannot be changed during runtime. Used for numbers and values which must be fixed (not changed) throughout the whole program.
	It is important to use this variable scope, when the same fixed value is used *more than once in the same PLC code.*
	See more in chapter 6.2, page 38.
RETAIN	Retains the variable value after a power failure or power loss. The variable is saved in memory (the internal memory).
	It is IMPORTANT to use this scope when a variable contains hour counters, items counters or similar, as these variable values must not be lost if PLC is (accidentally) turned off.
	See example in chapter 5.1, page 30.
	Cannot be used in a **FUNCTION**.
PERSISTENT	Similar to **RETAIN**. Variables are saved in an ASCII file on the hard disc.
	It is IMPORTANT to use this scope for values containing hour counters, items counters or similar. This is often only possible to use in a soft PLC.
	Contents of variables persisted (saved) on a hard drive are easy to move to other PLCs, e.g. if a PLC has to be changed.
	Cannot be used in a **FUNCTION**
END_VAR	End of the variable scope declare section
	Default (required)

5.1 EXAMPLE: Variables, Scope and IO-modules

This chapter shows an example with variable creation:

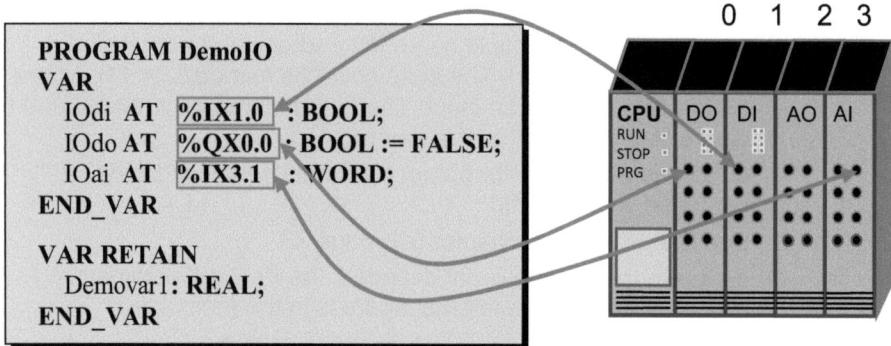

The example above shows four local variables in a program module named **DemoIO**.

There is a variable **IOdi** with the data type **BOOL** with a direct connection to port address no. 0 (first sensor input on the digital input card) on the hardware input module (PLC IO-card) no. 1. It does not make sense to initialize the variable as the value is determined by the sensor which is connected to the input card.

The output variable **IOdo** is by default set to **FALSE** to be sure that the output signal is set to zero (0 Voltage) when the PLC is turned on. It has a direct connection to port address no. 0 on the hardware output module no. 0 (the card closest to the CPU).

The input variable **IOai** is an analogue value with the data type **WORD**. An analogue input value card can be 16 bits but are typically 12 or 13 bits as these are cheaper and may provide sufficient resolution. The variable has a direct connection to port address no.1 on hardware output card no. 3

DemoIO has a local variable **Demovar1** with the data type **REAL**.

The local variable **Demovar1** is saved in memory in case of power failure or if power is turned off as it is marked with **RETAIN**.

Some PLC-types do not have a direct address on input or output card as shown above. In the PLC types, **%I*** and **%Q*** are written, as well as in a mapping table, a list of connection between variables and the physical input and output card, where it is possible to connect variables with the physical input and output card.

6 Naming the variables

Naming variables (tags) is an important task in PLC programming. This chapter and the following chapters contain guidelines and methods for naming variables.

Companies often have their own guidelines and conventions (rules) for variable naming, and the PLC programmer might have an opinion on how this should be done too. However, the most important rule to follow is to create meaningful variable names followed by a comment.

Variable names must begin with a letter after which the name can contain combinations of letters, numbers and some symbols, such as '_'.
Variable names must not be the same names as built-in functions, standard routines or user-defined functions. Variable names such as ARRAY, REAL or INT would therefore be invalid.

Variable naming rules & requirements:

- Invalid signs: ~ @ ; " # % & * : < > ? / \{ | },. SPACE, TAB.
- Invalid local language letters like Danish special signs: æøåÆØÅ.
- Use short indicative names: Some PLCs have a max character count of 24.
- Variable names cannot not start with a number.
- Do not to use the letter O close to a number (it can be mistaken for zero).
- The PLC does not distinguish between lower- and upper-case letters.

TIPS when naming with more words: First noun and then verb.

E.g. PumpRun, where Pump is a noun and Run is a verb
If a word has two nouns, begin naming with the big component:
E.g. **PumpSensorError** or **TankSensorLevel**

There are four methods of naming variables:

Hungarian Notation
Camel Case
Pascal Case
Snake Case

Hungarian Notation

Using this naming convention letters such as i, s, ar, b, are inserted at the beginning of the variable name to indicate which data type is used. However, if the variable is later changed to a different data type, this may cause issues in the code as variable names must be changed in both the PLC code and its documentation. Knowing the data type from the variable name might also be unnecessary as many programming tools today show the data type of a variable with a tool tip function (a small yellow box, appears if the computer mouse is held over the variable name).

Hungarian notation letters include:

X = BOOL, i = INT, I = REAL, ar = ARRAY, s = STRING, b = Bit, w= WORD, jw = DWORD, e= ENUM

Examples:	iMotorSpeed	(Speed on a motor with the data type **INT**)
	xMotorAlarm	(Alarm on the motor defined as the **BOOL** data type)
	sMotorAlarm	(**STRING** containing a motor alarm text)
	arMotors	(**ARRAY** with motors)

Camel Case

The naming convention Camel Case is where the variable name begins with a lower-case letter and the following words starts with an upper-case (capital) letter:

Examples:	flowMeasureWarningBit	blowerStartBit
	motorSpeed	calculateError
	sensorHighSignal	motorInitFunction
	sensorLow	powerEstimated

Pascal Case

The naming convention Pascal Case is where all words in the variable name starts with an upper-case letter.

Examples:	FlowMeasureWarningBit	BlowerStartBit
	MotorSpeed	CalculateError
	SensorHighSignal	MotorInitFunction
	SensorLow	PowerEstimated

This is probably the most commonly used method today, as it is easy to read, quick to write and creates the shortest variable name.

Snake case

This naming convention uses underscore to differentiate between words. Under-score is used as variable names cannot contain <SPACE>. However, this method can be difficult to read and names tend to become too long. Some PLC-types allow a maximum of 24 characters in a variable name which can become a challenge when variable names become very long.

Examples:	flow_measure_warning_bit	blower_start_bit
	timer_done_bit	calculate_error
	initial_motor_frequency	motor_init_function
	sensor_high_signal	power_estimated

A big advantage of using Snake Case is when tools for automatic generation of TAGS/variables are used in IO-Lists, electrical diagram drawings, PLC and SCADA codes, as "_" can easily be replaced with "." via the search-and-replace commands.

For abbreviations only <u>use</u> standard abbreviations such as **Cal** for calculate, **Avg** for average or **Cmd** for command.
If specific company or your own abbreviations are used, a comment must be written in the code or where the variable is created (defined), otherwise it might be difficult for readers of the code to figure out what the abbreviation means.

Below you see two identical PLC code examples where Pascal Code and Snake Case are used to create variable names. Consider which PLC code is the simplest to read (the one on the left or on the right):

```
IF TankLevel >= EmptyLevel THEN          IF tank_level >= empty_level THEN
    ValveOpen := TRUE;                       valve_open := TRUE;
    IF ValveError = TRUE THEN                IF valve_error = TRUE THEN
        ValveOpen := FALSE;                      valve_open := FALSE;
    END_IF;                                  END_IF;
END_IF;                                  END_IF;
```

Choosing between naming conventions

Which of the naming conventions is the best one is often a matter of opinion and is influenced by the methods the programmer has used before.

It is very important to choose meaningful names for variables. An example could be a variable which shows the status for pump no. 141. The pump could be given either of the following names:

Pump_Status_141, Status_141, P141_Status, Pump141Status, PumpStatus_141, P141S, osv.

Pump141Status is the best choice as the noun appears first in the name. The number of the pump (141), linked to the noun, appears after the noun (Pump) and finally the verb (Status). Furthermore, Pascal Case is chosen as the naming method because it creates short readable variable names.

Variable names including only one letter i, j, x, y, z, k, n are typically used for iterative variable (e.g. counters and loops) and index/pointers in **ARRAY**. It is faster to write a single letter than to write e.g. **ArrayIndex**. Often x, y and z are used in coordinate systems and 3D array.

Variable names such as **Temp1** and **Temp2** can be applied as temporary variables. These should not be used often, as they do not tell the reader what the variable is – i.e. the names are not meaningful.

Variables with names containing words such as **New** and **Changed** must be used carefully, because they are not new to the programmer who will be working on the PLC code later on.

Some programmers prefer to use the data type as part of the variable name, e.g. Int_Number_of_Run and Real_Initial_Temperature. This is somewhat similar to the Hungarian Notation but creates long names and can cause issues if the data type is changed later on.

Adding a unique number in front of each variable name makes the variables easier to identify and find in the code and documentation:

B8040_MotorSpeed	S213_PumpAlarm
B8041_MotorCurrent	S101_PumpSpeed
B8044_MotorPower	S001_SoftWareVersion
B9000_ValueToHMI	S501_ReadFieldbusData

We have now covered different ways of naming variables. The methods can, however, easily be used for the naming of functions, function blocks and program modules.

Some programmers write **fb** and **fc** in front of their own functions and function blocks:

fbCalculateArea	**fcArrayFindMin**
fcMotorStatus	**fcArrayFindMax**

Many built-in standard functions and routines do not use fb and fc in the names, which makes it difficult to be consistent.

Names such as B1, B2, B3, B4 etc. completely lack meaning and should not be used unless they follow logic used in the problem statement (the control requirement specification or functional description). The names are used to reduce the page space requirements of the book.

When naming a **STRUCT** (see chapter 4.4, page 20) *TYPE* can advantageously be added to the name to make it is easy to see that **STRUCT** is used.

Alarm texts can be of both the data type **STRING** consisting of text or **INT** for alarm numbers. Multiple languages can be displayed on the control panel at the same time:

sAlarmMotorLoad_DK	"Alarm motor overbelastet"
sAlarmMotorLoad_UK	"Alarm motor overload"
iAlarmMotorLoad	12004

Many companies within the process industry (dairies, breweries, the medical industry, the oil industry) follow the S88 naming standard (ANSI/ISA-88). This standard is focused on the sensor type and the installation location when creating names. The standard covers the IO-list, the control specification, the PLC program and test documents. Using the same naming standard of the variables in the entire control system obviously creates fewer misunderstandings and better quality, as well as making it easier to have a good overview of all variables, the PLC code and the documentation.

Examples:

FZ.MM01.UE01.PO3	FZ_MM01_UE01_PO3
FZ.MM02.UE01.M01	FZ_MM02_UE01_M01
FZ.MM02.UE01.TT01	FZ_MM02_UE01_TT01

In the example above **PO3** is the "Control Module", **UE01** is the "Equipment Module" and **MM01 is** the "Process Cell" according to the S88 naming standard.

6.1 Variables with unit of measurement

It is often necessary to connect a variable to a unit of measurement. If for example a variable is used to represent a temperature, the temperature must be stated in °C (degrees Centigrade/Celsius) or °F (degrees Fahrenheit).

It will help the programmer if the unit is added to the variable name to make it visible during the programming phase. A variable measuring temperature in degrees Centigrade can be named **MeasureTemperatureC**, where C indicates the unit. Also write the unit in the commentary field where the variable is created.

Examples of other variables needing a unit of measurement:

Variable	Potential units
Time, period [#1)]	us, s, seconds, minutes, hours, days, week, year
Speed	m/s, km/t, km/h, rpm, %, mph, mm/s, tf/s
Amount	kg, g, No., DKK, dollars, pcs., liters, bottle, box
Weight	kg, pounds, lbs., g, tons, mg, %
Oxygen	mg/l, %, g, l
Consumption	W, kWh, Dkr, l, kg, $, m, m2, m3, A, k/j, g, l/h

It is important to obtain an overview of all measurement units. In some PLC Control Systems, it is required that the PLC Control System itself is able to change/convert units, especially if the same PLC Control system is used globally. E.g. indicating whether a temperature is measured in °C or °F. This is typically the case when development PLC Control Systems for the US and Canadian markets, where it may be required to implement functionality where temperature units can be changed online.

You can find conversion formulas online. Below example shows how to convert °C to °F in a PLC program:

```
VarF:= (VarC * 9/5) + 32;
```

Units can be SI-units (m, kg, s, A); pay attention to the SI-prefix.

When displaying units on the human-machine interface (HMI), in data logs for files and reports, etc. it is often required that values do not exceed two decimal points. On a HMI, values with two decimal points are often displayed with %f5.2 in the text field. The %f means a FLOAT (REAL) value and 2 means two digits after the dot as shown:

$$23.45 \; [^\circ C]$$

Units are often written with square brackets to increase the readability; e.g. temperature [°C]. Use this both in the HMI and its documentation.

It makes sense to create one piece of reusable code to convert values from one temperature unit to another. You can re-use this piece of code across PLCs and customers. An example of this is shown in chapter 10.4.1, page 87.

Note all PLC types save comments in the PLC, and when uploading the PLC code to a PC the comments in variables fields can be lost. If the unit is part of the variable name the units are not lost.

NOTE: Time, period #1)
Time can be difficult to work with when programming for international use. There can be differences in which day of the month the countries shift between summer and winter time, and whether Sundays are the first or last day in a week. Finally, there are differences in when week number 1 in the calendar year starts.

Examples of variables with units as part of their names:

```
TemperatureC
TemperatureF
MotorSpeedHz
MotorSpeedPercent
ConsumptionW
ConsumptionKWH
MotorUseA
```

Note that names may not contain the signs: % (percent), / (slash) and ° (degree).

6.2   Variables with fixed values (CONSTANT)

Variables, fixed and unchangeable during the program execution, must be configured as a constant value (**CONSTANT**).They are used for numbers used more than once in the same PLC code. This ensures the same value is used everywhere in the code.

CONSTANT variable names are often written in CAPITAL letters (uppercase).

When must a CONSTANT be used?
If a value is used several times, e.g. 25.4, which is the converting factor between millimeters and inches, a constant must be defined to be used in the code:

$$MILLI\text{-}METERS\_PER\_INCH = 25.4.$$

However, as it is not likely that the converting factor between millimeters and inches needs changing, the programmer might prefer to use the number 25.4 rather than a long variable name. In case the number would need changing, this could easily be done with a 'search and replace' command. However, 'search and replace' commands can be dangerous as it can make unintended value changes. Using **CONSTANT** when naming variables creates safe and stable programs. Furthermore, if constants are used, they will also contribute to a self-explanatory program, because text is used instead of just a number e.g. 25.4

When declaring and creating an **ARRAY**, the length must always be defined as a **CONSTANT**. This is due to the risk of making the program unstable if the length value is not changed throughout the code when the programmer carries out potential updates to the **ARRAY**. The length of an **ARRAY** is changed when e.g. testing the **ARRAY**. See an example chapter 9.4.3, page 74, where **BufArrayMin** and **BufArrayMax** are created as constants and used with an array named **BufArray**. By adding **Min** and **Max** to the array name, it is clear to see that they belong together:

Benefits of using constants:
- The PLC Code is more readable
- Avoid errors when changing constants and values
- Save time when changing a value

Examples of using constants

PI:=	3.1415927
SECONDS_DAY:=	86400.0
ARRAY_MAX:=	10

If an integer value is used in a calculation with real numbers, write the integer with a zero digit e.g. 3.0 in order to ensure that the CPU handles the calculation correctly.

7 Operators, MATH and LOGIC

The following chapters describes the arithmetic, logic and relational operators used in PLC programming.

A PLC has the same built-in math functions as known from a regular calculator.

7.1 Arithmetic Operators (+, -, *, /)

Table of arithmetic operators (mathematical symbols):

Operator	Explanation	Function *)	Examples where V1 = 2 V2 = 5	? Y =
+	Addition	Y:= **ADD**(V1,V2);	Y:= V1 + V2;	7
-	Subtract	Y:= **SUB**(V1,V2);	Y:= V1 – V2;	-3
*****	Multiply	Y:= **MUL**(V1,V2);	Y:= V1 * V2;	10
******	Exponent	Y:= **EXPT**(V1,V2);	Y:= V1 ** V2;	32
/	Divide	Y:= **DIV**(V1,V2);	Y:= V1 / V2;	0,4
MOD	Modulo	Y:= **MOD**(V1,V2);	Y:= V2 **MOD** V1;	1

V1, V2, Y can be numbers (integer or decimal numbers) or variables.

***)** The functions are normally used in LD programming and not all PLC types support the functions in ST programming.

The built-in functions **ADD**, **SUB**, **MUL**, **EXPT** and **DIV** from the LD programming can be used. But in ST-programming it does, however, make better sense to use the arithmetic operators (see table above) as it reads like 'ordinary' calculations.

Not all PLC types support the ** operator. Instead use the EXPT function:

Example: $C=(2^a-b)*2$ => C:=(2**a-b)*a; => C:=(**EXPT**(2,a)-b)*a;

One of the strengths of ST programming is that math calculations are similar to the methods used in math-programs and consequently the calculations are simple to write, troubleshoot/debug and read in the PLC code.

Examples of math operations can be seen on page 45 and page 132.

To perform calculations, it is important to choose the right data types for the variables. In most cases, a **REAL** variable will be the right data type.
If e.g. **INT** is used as a data type, the calculation can in some cases create a variable overflow as the data type is too small and cannot contain the size of the result of the calculation. This is due to the fact that the calculation results in a larger number than can be contained in the chosen data type. See also chapter 8.3, page 49.

This can be illustrated in the following example:

Calculating: **Y = V1**V2**, $(Y = V1^{V2})$

where V1 = 10 og V2 er 10, result:

Y = 10000000000

The value Y is to large for an **INT** data type

IMPORTANT
Choose the right data type for the calculation.
If too large a data type like **LREAL** or **LWORD** is chosen, more memory is used and it requires longer scan time than necessary.

7.2 Relational Operators (=, <, <=, >, >=, <>)

To compare the relation between two values (integer or decimal numbers) use relational operators. The two values can be variables or numbers.
The result of the comparison is a value, which always has the data type Boolean (**BOOL**) and can therefore only be **TRUE** or **FALSE**.

The relational operators are:

Operator	Description
=	Equal
<	Less than
<=	Less or equal
>	Greater than
>=	Greater than or equal
<>	Not equal

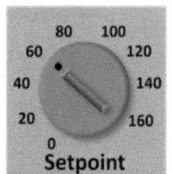

Example of use:

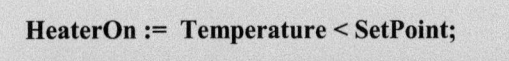

HeaterOn := Temperature < SetPoint;

Hand turning knob

The data types for Temperature and **SetPoint** are both **REAL**. The expression can be used if e.g. a heat lamp has to be switched on if the temperature is too low.
Temperature can be measured by a sensor connected to an analogue input module.

The **SetPoint** variable contains the temperature at which the heat lamp should turn on, and the value could come from a hand turning knob with a potentiometer (see figure).

Explanation:

HeaterOn will be **TRUE** if Temperature is lower than **SetPoint**. As the expression **Temperature** < **SetPoint** results in a variable of the data type **BOOL**, **HeaterOn** must be a **BOOL** data type. The variable **HeaterOn** can be connected to a digital output module, which when **TRUE**, activates a relay that turns on the connected heat lamp.

Relational operators are mostly used in **IF**-statements, see chapter 9.1, page 56.

7.3 Numeric Operators (MATH functions)

This chapter describes the built-in math functions in a PLC.

Math functions have typically only one input parameter - a number of data type **INT** or **REAL**. The return parameter from the function is often of the data type **REAL**. It is important to ensure that the input parameter is valid. It is e.g. not possible to call the **LN** function with the value 0 as this is not mathematically correct and as a result the PLC Controller will stop the program execution (Run Time Error).

A correct program execution can be carried out as follows, where **x** is an input parameter and **y** is the result when calling the **LN** function:

```
IF x <> 0 THEN
    Y = LN(x); //Only calculate if x is not zero
END_IF;
```

Below table shows a list of built-in math functions in a PLC:

Function	Mode of operation (Example where a = 2, b = 5, c = 8)
NEG	Change a positive number to a negative number and vice versa. Same as **a:= a * -1;**
INC	Increases by 1. Add 1 to the value. Increment, **INC**(a) = 3. The same as **a:= a + 1;**
DEC	Decreases by 1 down, Decrement. **DEC**(a) = 1. The same as **a:= a - 1;**
TRUNC	Converting a **REAL** value to an **INT** value. The integer value does not get rounded, instead values after the dot are removed. **TRUNC**(3.9) = 3 **TRUNC**(-2.5) = -2 The function removes the digits after the dot.
FRAC	The decimal value of a **REAL** value. **FRAC**(2.8) = 0.8, **FRAC**(-3.49) = 0.49
ABS	Absolute value. The function ensures a positive value. **ABS** (-1.2) = 1.2 **ABS** (3.4) = 3.4 **ABS** (-3) = 3
FLOOR	For positive values, the return value is less than or equal to the input For negative values, the return value is greater than or equal to the input. **FLOOR**(2.8) = 2 **FLOOR**(-2.8) = -3

Function	Mode of operation (Example where a = 2, b = 5, c = 8)
SQR	Square. This function calculates x^2, raised to the power of 2. **SQR**(4) = 16, The same as x * x, (x multiplied by x).
SQRT	This function calculates the square root. **SQRT** (4) = 2, **SQRT**(9) = 3
LN	The natural logarithm. **LN**(2.71828) $\approx$ 1 ((the wave sign means approx.).
LOG	The natural logarithm with base 10. **LOG**(10) = 1.
EXP	Exponential function. Same as e^x or e^x, e = 2.718281828 **EXP** (1) = 2.718281828
SIN	Sinus funvtion. **SIN**(a) = 0.35 (GRAD) [#1)].
COS	Cosinus funktion. **COS** (a) = 0.99939 (GRAD) [#1)].
TAN	Tangent function. **TAN**(a) = 0.03492 (GRAD) [#1)].
ASIN	Arcsin function. Inverse sinus function. $SIN^{-1}(x)$, Sinh(x) [#1)].
ACOS	Arccos function. Invers cosinus function $COS^{-1}(x)$, cosh(x) [#1)].
ATAN	Arctan function. Invers tangent function $TAN^{-1}(x)$, tanh(x) [#1)].
EXPT	Exponentiation of a variable with another variable. a^b = **EXPT** (a,b) = 2^5 = 32

The above usually appears as built-in functions in a PLC; i.e. functions which can be used without creating extra program library (add-ons) or program modules. Small variations in the functions can exist between the different PLC-types. Always look through the programming manual of the PLC-type to gain an overview and see the possibilities for math functions and routines.

To use the right data type, remember to check the variable data type for each individual math function.

#1) To calculate between radians (RAD) and degrees (GRAD) see page 51.

7.4 Logic Operators (AND, OR, XOR, NOT)

Logic operators are used to compare two different **BOOL** variables or values. The result of the comparison is a value, which always has the data type Boolean (**BOOL**) and can therefore only be **TRUE** or **FALSE**.

See below for operators and examples:

Operator	Description	Example S1:= TRUE, S2:= FALSE S3:= TRUE	Result
&	Same as **AND**, only **TRUE** if both values are **TRUE**	K1:= S1 **&** S2 K2:= S1 **&** S3	K1 = **FALSE** K2 = **TRUE**
AND	AND, result is **TRUE** if both values are **TRUE**	K1:= S1 **AND** S3 K2:= S1 **AND** S2	K1 = **TRUE** K2 = **FALSE**
OR	OR. **TRUE** if one value is **TRUE**	K1:= S1 **OR** S2 K2:= S1 **AND** S3	K1 = **TRUE** K2 = **TRUE**
XOR	The result is **TRUE** if the values are not equal.	K1:= S1 **XOR** S2 K2:= S1 **XOR** S3	K1 = **TRUE** K2 = **FALSE**
NOT	not, negated **TRUE** result if value is **FALSE** **FALSE** result if value is **TRUE**	K1:= S1 **AND NOT** S2 K2:= **NOT** S1 K3:= **NOT** S2	K1 = **TRUE** K2 = **FALSE** K3 = **TRUE**

Logic operators are mostly used with **IF**-statements as described on page 56.

AND can be used in serial-connected components (sensors/contacts/switches), where all components must provide an ON signal to make the entire expression **TRUE**.
OR can be used in parallel-connected components, where just one component is needed to provide an ON signal to make the entire expression **TRUE**.

The logic operators can also be used directly on e.g. binary values as shown below:

```
Var1 := 2#10010011 AND 2#10001010;    // Var1 er 2#10000010
```

```
Var2 := Var1 OR 2#10001010;    // Var2 er 2#10001010, DEC138
```

7.5   Logic, math formulas and use of parentheses ()

It is important to be aware of how math formulas are calculated in a PLC Controller. If in doubt of how values are calculated – if addition comes before multiplication – use parentheses.

> Following mathematical rules *multiplication* is carried out before *addition*, but experience shows that you cannot be 100 % sure that the rules of math are respected in a PLC or that the formula is written correctly in the PLC Code. Therefore, use parentheses to be sure

If the math formula contains Boolean expressions like **AND** or **OR** as shown below :

```
X:= B1 OR B2 AND B3;
```

Then **AND** is read as 'multiply' and is calculated first. **OR** is read as 'plus'.

It means: if the value of **B2** is **FALSE**, then the expression **B2 AND B3** is **FALSE**.

If you are in doubt about the result use parentheses as shown below:

```
X:= B1 OR (B2 AND B3);
```

The next example is this formula:

$$V1 = \frac{V2}{V3} + \sqrt{(V4 + V5)}$$

The formula can be written in the PLC code with extra parentheses as follows:

```
V1: = (V2/V3) + (SQRT(V4 + V5));
```

SQRT is the mathematical function in a PLC calculating a square root. The function only has one input parameter. **V4** and **V5** are added before calling the function.

8 Variable assignment

Variables are an important part of programming. In this chapter, the basics of variable assignment will be covered along with important information and tips when working. Variables are in some PLC types called **tags** or **PLC tags**.

> **Definition:** A variable points (link) at a box in the memory containing a place in which a numerical value can be written. The size of the box depends on the data type which is very important to remember.

The example below shows how the variable with the name **VarA** gets a copy of the number which exists in the variable **VarB**. That means **VarA** is assigned the value held in **VarB**. Note the use of the signs : = and ; (colon, equal and semi-colon).

> **VarA:= VarB;**

Subsequently **VarB** can be given a value of 17.6 as follows:

> **VarB:= 17.6;**

Dot (.) is always used in a PLC when decimal numbers are applied. Both the variable **VarB** and **VarA** has the data type **REAL** (**REAL** is used for decimal numbers).

If the data type for **VarB** is an **INT** (integer) it is common that the compiler (the program in which the PLC code is written) comes up with a warning message telling the programmer that values may be lost, as the number assigned to **VarB** is a decimal number (17.6). This is due to the fact that the variable **VarB** can only contain an integer if it is created with the data type **INT** (integer).

In ST-programming working with variables and calculations is very simple.

This calculation is written directly in a PLC code:

> **VarB:= 17.6 * 8 + VarA;**

If the value of the variable **VarA** is 23, then the value of **VarB** is 163.8

The variable **Count** shown below will at each program-scan be increased by 1 (1 is added to the previous value). The program execution has an internal variable for calculations (called Stack/Accumulator) which makes a copy of the variable **Count**, adds 1 to it and returns the new value to **Count**:

```
Count:= Count + 1;
```

If **Count** is of the data type **INT**, be aware that when **Count** reaches the value 32767, it will change to – 32768 next time the program runs (next program scan). It is the programmer's responsibility to make sure that no overrun happens on a variable. There are two methods to do this: Either a larger variable is used for **Count** e.g. **DINT**. Or the counter is created with a condition (**IF**-statement, see chapter 9.1, page 56), setting the value to 0 when the number reaches the maximum value.

The last method is the best one as it prevents variable overruns:

```
Count:= Count + 1;

IF Count > 99 THEN //To avoid overrun
   Count:= 0; //Reset counter
END_IF;
```

As shown above, **Count** adds 1 with each program-scan. If the PLC scan time is set to 1 [ms] it will take 100 [ms] before **Count** is reset to 0.

TIP: The above counter can easily be used as a program *Heartbeat*, making it possible to see activity on a running PLC program.

The following are built-in counting functions: **CTU**, **CTD** and **CTUD**. See page 112.

8.1   MATH calculations challenge

Where formulas are involved, math and calculations are easy to work with in ST-programming. This is one of the biggest advantages compared to the other PLC programming languages. However, there are several things which should be taken into consideration when working with math functions and formulas. These are:

- Division by 0 (chapter 8.2, page 48)
- Calculating with **INT** and **REAL** (chapter 8.3, page 49)
- Decimal errors when using **REAL** (chapter 8.4, page 50)

8.2   Division by zero

It is important for the PLC programmer to remember that a PLC reads data from different sensors, and some of this data may contain output values of zero. For example a thermometer (temperature sensor with a transmitter) measuring temperatures outside can have a 0 degree value output. This is shown in the calculation below, where **VarC** is equal to **VarA** divided by **Temperature**:

```
VarC:= VarA / Temperature;
```

If **Temperature** becomes zero, the PLC will produce a run time error and/or become unstable because it is an invalid mathematical operation in a PLC.

To ensure that the PLC does not produce a run time error at any time and to minimize the risk of errors occurring later, the above PLC code can be changed as follows:

```
//Ensure temperature is not zero when calculating
IF Temperature <> 0 THEN
   VarC:= VarA / Temperature;
END_IF;
```

The calculation is only carried out if **Temperature** is not zero (the operator sign **<>** means different from / not equal. See chapter 7.2, page 41).

Another possibility to ensure that the calculation is not carried out when the **Temperature** is zero, is the following solution:

```
//Ensure temperature is not zero when calculating
IF Temperature = 0 THEN
   Temperature:= 0.0001;
END_IF;

VarC:= VarA / Temperature;
```

NOTE: The math functions **LN** (x) and **LOG** (x) cannot tolerate **x** with a value of zero.

8.3   Calculating with REAL and INT variables

Calculations can be made with both integers (**INT**) and decimal values (**REAL**). If a division of two integers is to be calculated it must be considered which data type the variable is using and how the calculation is carried out.

In the example to the right the three variables are all of the type **INT**, where the brown box shows the value inside the variable:

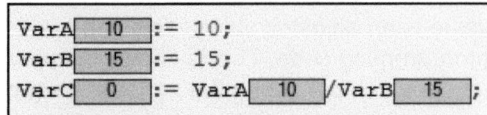

In the example above the result of **VarC** will be zero because **VarC** is an integer. In an ordinary math calculation this example should provide a result with decimal values (10 divided by 15 does not result in an integer, but the decimal number 0.67). However, as the division is carried out with values of the data type **INT** the result is zero.

To make the calculation succeed, the calculation has to take place in a **REAL** variable. The calculation inside a PLC is carried out by using the data type for the calculation which *is the first variable* in the formula/calculation. In this example it is the data type for **VarA**. The PLC will disregard the data type of **VarC** which is in fact a **REAL**. However, **VarC** must be a **REAL** for the result to be saved (it is not possible to save a **REAL** variable in an **INT** variable).

As **VarC** is of the **REAL** data type, the solution to ensure a correct calculation is to copy **VarA** to **VarC** before the calculation is carried out

The internal calculation is then performed in a **REAL** variable. The code for this is as shown to the right:

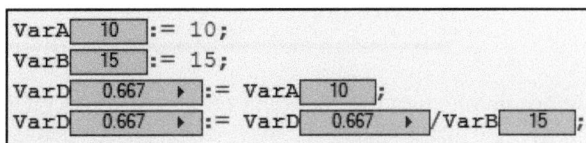

If numbers are written directly (hard coded) into a calculation, the PLC sees these as integers and the result is not as expected. **VarD** will become zero:

The solution is that numbers must be written as decimal numbers with a dot to ensure the calculation is performed correctly as shown:

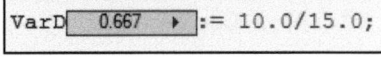

As calculations in a PLC can be different to what we know from our calculators, a tip is therefore to check whether the calculation shows the expected result. A calculator or a math program must be used to make control calculations.

In some PLC types, a calculation is made inside an accumulator (ACC) from where values must be copied to and from. The same rules apply.

8.4 Decimal errors when using REAL

When calculations are made using the **REAL** data type, it sometimes occurs that a result is not a round number. A variable result might be expected to be a nice, round number such as e.g. 11, but the calculation results in the number 10.999999. This is caused by the fact that a computer can only work with the integer data type. A **REAL** value is an adjusted value. This can create problems when comparing numbers in the programming code. This is shown in the code lines below where the variable **Lamp1** must be set to 1, when the variable **Sensor1** becomes 11

```
IF Sensor1 = 11 THEN
   Lamp1:= TRUE;
END_IF;
```

Because a decimal error might occur which has the consequence that Sensor1 never becomes exactly 11, there is no guarantee that the above PLC code will work correctly and set the variable **Lamp1** to **TRUE**.

The above example can be changed to the below PLC code where the **Sensor1** value is now checked inside a range. The range could be between 10.99 and 11.01:

```
IF (Sensor1 > 10.99) AND (Sensor1 < 11.01) THEN
   Lamp1:= TRUE;
END_IF;
```

Alternatively, use the rounding off functions **FLOOR**() or **TRUNC**(). See page 42.

You can also implement a rounding function (here the rounding is set to 1 decimal):

1) Multiply the sensor value by 10
2) Convert value to an **INT** variable by using the function: **REAL_TO_INT**();
3) Convert back to a **REAL** variable by using the function **INT_TO_REAL**();
4) Divide value by 10

Problems with rounding can be experienced in a number of different contexts. For example, when a motor's speed gauge is close to but not 0 (zero), or a fuel tank is physically empty but the level sensor shows a small value close to zero. A flow meter can also show a small value even if the plant is not in operation. This can, however, be caused by a lack of calibration (zero position) of the instrument.

8.5   Data communication (transfer of variables)

When designing automation solutions variables often needs to be transferred to other computers. This chapter highlights issues in relation to data transfer of variables.

Problems can occur when transferring **REAL** variables to other PLCs, PCs, electrical apparatuses or automation instruments. This happens due to different interpretations of how a **REAL** or FLOAT value is defined in different computers (a computer represents works in integers more precisely). Issues can also arise due to different programming versions, or that 16, 32, 64, 128-bit systems handle **REAL** and FLOAT differently. This problem is solved by always transferring values as integers. Values can then be multiplied by 100 to obtain values with two decimals and the receiver must then divide by 100 to obtain the right decimal values with two decimals.

Due to different ways of handling and interpreting **STRINGS**, it can be a challenge to transfer **STRINGS** between computers. Issues could occur due to different bit sizes, Unicode or choice of **ASCII** characters. Lastly, the length of a **STRING** starts at position zero in some programming languages and position 1 in others. Converting **STRING** to **BYTE** makes it 'simple' to transfer data.

Always start data communication by reading **WORD**. Remember that some computers have swapped **WORD** values (the lowest 8 bits are replaced with the 8 highest bits). Also remember that if a value begins with 0X, it is a HEX value.

In some PLCs, a **BOOL** uses 16 bits and can therefore also be an **INT**.

A **STRUCT** cannot be transferred directly as this is a structure (a kind of template). If a PLC is to receive a **BOOL** variable (alarm signal, counter value, trigger signal, or start signal), it must often be implemented as a one-shot in the PLC.

All variables passed between several devices can be advantageously noted in a protocol description. A kind of I/O list, so there is documentation for the communication.

.

8.6   Data type conversion functions

Several built-in functions are available to transfer (converting) the value of a variable with one data type to a variable with another data type. Some PLC types supply more than 100 different conversion functions for different data types.

Naming and syntax for the conversion functions:

> **TYPE1_TO_TYPE2** (ConvertFrom);

Where

> **TYPE1** is the data type which is being replaced by the
> data type of Type2 (data type of ConvertFrom).
> **TYPE2** is the converted data type

Most used datatype conversion functions:

Function	from	to	Example	Comments
REAL_TO_INT	REAL	INT	Val:= **REAL_TO_INT**(1.6); \\Val = 2 Val:= **REAL_TO_INT**(1.3); \\Val = 1	Rounding to nearest integer (IEC60559) **Val** is an **INT**
INT_TO_REAL	INT	REAL	Val1:= **INT_TO_REAL**(4); \\Val1 = 4.0	Convert an integer to a decimal value. Val1 is a **REAL**
INT_TO_BOOL	INT	BOOL	Val2:= **INT_TO_BOOL** (1); \\Val2 = TRUE	1 is converted to **TRUE**. 0 is converted to **FALSE**.
INT_TO_TIME	INT	TIME	Val3:= **INT_TO_TIME** (5); \\Val3= T#5ms	Converts an integer value to a variable with the **TIME** data type in [ms] See note **#1)**
RAD_TO_DEG DEG_TO_RAD	LREAD	LREAD		Converts between radians (RAD) and degrees (GRAD). Used with the **SIN** and **COS** functions.

To convert a **REAL** variable (decimal value) to an **INT** variable, the function **REAL_TO_INT** must be used. See first row in the above table.

It is important to ensure that the value *can* be converted, because if it can't an error can occur causing the PLC to stop the program execution, or make the entire program unstable.

#1) DATE is converted from an internal electronic circuit which is a part of the hardware in a PLC. This circuit counts time in seconds from 00:00:00 UTC 1.1.1970 (Coordinated universal time, atomic clock).

Note that the next Y2K will occur in the year 2038.

8.7   Finding binary values of an integer (Masking bit)

Sometimes there is a need to convert an integer value into a binary value to check whether a specific bit in a variable is **TRUE**. This is typically needed when different digitals output (e.g. lamps) is set from an integer value.
This is also called: To mask out the binary digit from an integer.

This can be carried out in a simple way: Use dot and a digit (bit position no. 0) after the variable as shown below:

```
MyUINT:= 3;     //Unsigned INT datatype. The BIN value is 2#0011
MyBOOL2:= MyUNIT.0; //Get bit 0 from MyUNIT variable
```

MyBOOL2 (**BOOL** data type) is **TRUE** because position 0 in **MyUINT** is the first bit, which is 1 in a value that is 3.

The above can also be written as follows:

```
MyUINT4:= MyUINT AND 2#001;  //Where MyUINT = 3 = 2#0011
MyBOOL:= UINT_TO_BOOL (MyUINT4); // Convert to a BOOL
//The result is that MyBOOL is TRUE
```

Where **AND** can be used to mask out a bit at position no. 0. Each bit in the two values **MyUINT** and **2#001** are 'multiplied binary', and if the result is 1, the final result will be **TRUE**. When '2#' is placed before a value, it means that the value must be used as a binary digit. See also chapter 4.1, page 14.

If the result needs to be a variable with a **BOOL** data type, the conversion function **UINT_TO_BOOL** must be used.

Below is an example where a variable named **Var1** (**UINT**) is used to set different outputs bits. The variable **OutPutBitX** will be set to **TRUE** conditions is **TRUE**:

```
OutPutBit1:= Var1 = 2#00001; //TRUE if Var1 = 1
OutPutBit2:= Var1 = 2#00010; //TRUE if Var1 = 2
OutPutBit3:= Var1 = 2#00011; //TRUE if Var1 = 3
```

8.8   Valve matrix

This example shows how variables can be used to control a valve matrix. A valve matrix is used in breweries and dairies, where several tanks must be emptied into one or more shared pipelines. The example here uses a valve matrix designed for 5 tanks which can distribute liquid into 3 different pipelines, as shown in the picture:

An obvious solution is to declare a 2D array with a **BOOL** datatype, where each **BOOL** variable is directly connected to the valve in the matrix:

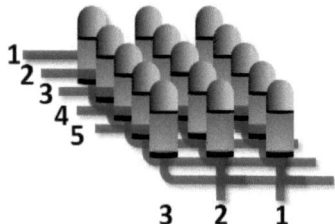

```
VAR PipeMatrix
  ARRAY [1 ..5, 1 .. 3] OF BOOL;
END_VAR
```

Here, tank 3's valve is open and the liquid flows to pipeline 2:

```
PipeMatrix [3,2]:= TRUE; //Empty tank 3 to pipe 2
```

However, creating an array of **UINTs** may provide a better overview as shown below:

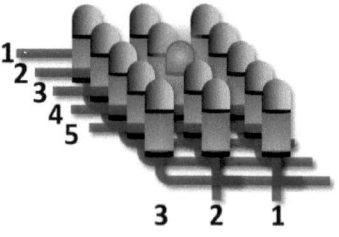

```
VAR PipeMatrixUINT
  ARRAY [1 .. 5] OF UINT;
END_VAR
```

Opening a valve is done by typing in 1 at Tank 3, as shown (2 # = binary number):

```
PipeMatrixUINT[1]:= 2#000; //Tank 1
PipeMatrixUINT[2]:= 2#000; //Tank 2
PipeMatrixUINT[3]:= 2#010; //Tank 3
PipeMatrixUINT[4]:= 2#000; //Tank 4
PipeMatrixUINT[5]:= 2#000; //Tank 5
```

It is the valve connected to bit 2 in array 3 that opens and the rest of the valves are closed. The method provides a good overview of all valves.

All bits can be extracted and copied into **PipeMatrix** in this way (see more page 53).

Note: Only two valves are shown below. There will be 15 lines of code in total:

```
PipeMatrix[1,1] := PipeMatrixUINT[1] = 2#001;
PipeMatrix[1,2] := PipeMatrixUINT[1] = 2#010;
//The statement " PipeMatrixUINT[3] = 2#XXX" is TRUE or FALSE
```

8.9   Rounding a REAL to 2 decimals (2 digit REAL)

If a **REAL** value is converted into a **STRING** and read out on an HMI (user interface) or written to an ASCII file, the value will often include 7 to 9 digits. That many digits are not very readable and user-friendly. It is, however, a way for a computer to handle a decimal digit. A **LREAL** data type, for example, has 15 digits.

The method below rounds the value in **RealNumber** to a digit with two decimals. If three decimals are required the constant **DecimalFactor** must be 1000:

```
VAR CONSTANT
    DecimalFactor : REAL := 100; //10 for 1 digits, 100 for 2 digits, 1000 for 3 digits
    RealNumberBegin : REAL := 50.7172;
END_VAR
VAR
    INTNumber: INT;   RealNumber: REAL ;
END_VAR

RealNumber:= RealNumberBegin;
IF DecimalFactor > 0 THEN //Avoid division by zero (0)
    RealNumber:= (RealNumber * DecimalFactor) + 0.5; //+ 0.5 to round up  #1)
    INTNumber:= REAL_TO_INT(RealNumber);     //  Convert to integer  #2)
    RealNumber:= INT_TO_REAL(INTNumber);     //  Convert to decimal #3)
    RealNumber:= RealNumber/DecimalFactor;   //  Add decimal         #4)
END_IF;
```

DecimalFactor is a **CONSTANT** because is used more than once in the PLC code.

A calculation example, where 50.7175 is converted to 50.72:

#1) (50,7175 * 100) + 0,5 = 5072,25
#2) REAL_TO_INT (5072,25) = 5072 (integer value)
#3) INT_TO__REAL(5072) = 5072 (decimal value)
#4) 5072/100 = **50,72**

IMPORTANT:

Rounding must not be carried out before other calculations as it deletes information. Only perform rounding when the value is shown to the user:

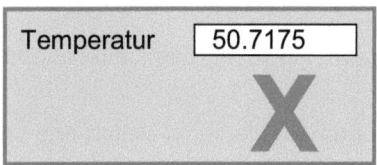
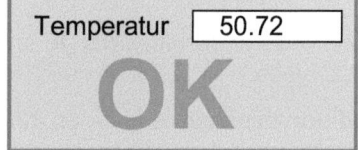

9 Basic ST programming

The following chapters describe the main declarations and concepts in ST. This includes the basic ST programming syntax followed by a number of code examples. Syntax means the set of rules, principles and structure of programming.

In the shown syntax descriptions, the **<Condition>** and **<Statement>** sections will be replaced by variables, logic, expressions and PLC code.

9.1 IF-THEN-ELSE statement

An **IF-THEN-ELSE** statement is the most used expression in the ST programming.

The **IF** statement can e.g. be used to read a digital signal from a sensor input module and then perform an action. The sensor input can be an electrical start switch, an ON/OFF switch or a level switch in a pump well. If the sensor is activated, an action must be performed: e.g. starting a pump or turning on a light.

The **IF** statement can also be used for analogue input signals and for internal variables.

The syntax of the **IF** statement is:

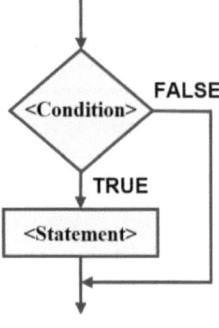

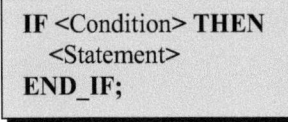

IF-THEN flowchart

Where:

<Statement> =	Can contain one or more lines of PLC code.

<Condition> =	An expression is always either **TRUE** or **FALSE**. If the expression is fulfilled, the code in **<Statement>** is carried out.

The syntax always ends with **END_IF** and semicolon. However, semicolon is optional in some PLC types.

The **<Condition>** line can e.g. be an input signal from an electrical switch or a sensor and the **<Statement>** line can be an output signal to turn a lamp on or off.

The **ELSE** statement can be added to the expression:

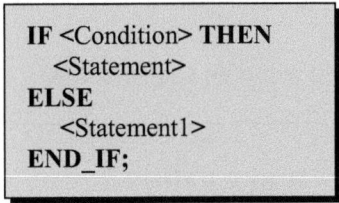

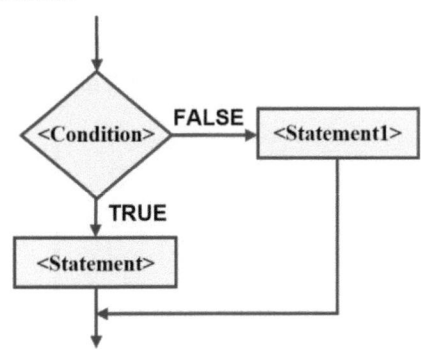

The **ELSE** statement is optional, and notice that the lines including **<Statement>** are indented (2 x space) to make the whole expression more readable.

*Flowchart of the **IF**-**ELSE** statement*

The mode of operation is as follows:

If **<condition>** is fulfilled (**TRUE**), the PLC code in **<Statement>** will be carried out.

If **<condition>** is *not* fulfilled (**FALSE**), the PLC code in **<Statement1>** is carried out.

The above syntax can be changed, so **ELSE** is not used:

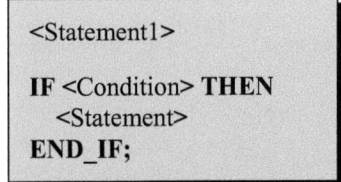

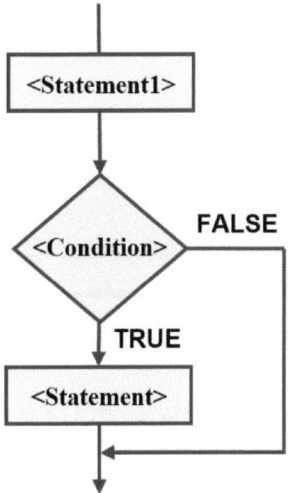

The mode of operation:

The code in the **<Statement1>** section will be executed first, which can be assignment of values to variables.
If **<Condition>** is fulfilled (**TRUE**), the code in the **<Statement>** section is executed and can be used to overwrite the assigned variables in the **<Statement1>** section.

*Flowchart without **ELSE** statement*

As seen in this example, it is not necessary to use the **ELSE** statement in an **IF** statement. Not using **ELSE** can make the code more readable.

NOTICE: If the **<Condition>** section contains colon "**:=**" the program will check to see whether the variable has been assigned successfully. However, this is not normally the intention!

Therefore, remember that the equal sign "**=**" must stand alone as shown below:

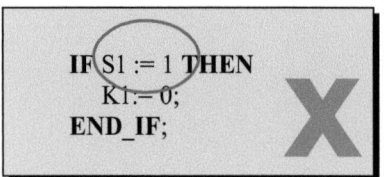

When using **ELSE** statements you can easily end up writing more complex code as shown below:

```
IF <Condition1> THEN
    <Statement1 >
    IF <Condition2> THEN
        <Statement2>
    ELSE
        <Statement3>
    END_IF;
ELSE
    <Statement4>
END_IF;
```

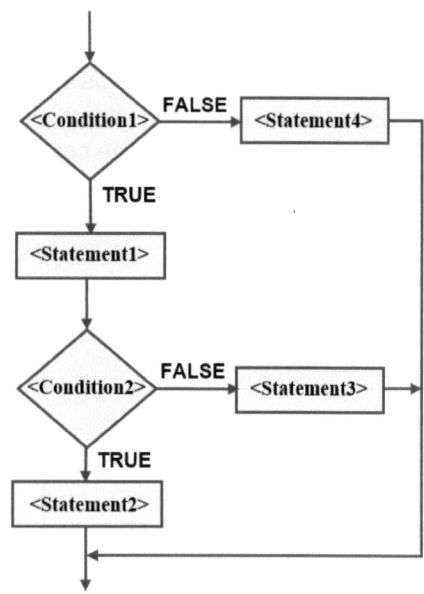

The mode of operation is as follows:

The PLC code in **<Statement1>** is executed, if **<Condition1>** is fulfilled (**TRUE**). After **<Statement1>** has been carried out, **<Condition2>** is controlled and if it is **TRUE**, **<Statement2>** is carried out; otherwise **<Statement3>** is carried out. If **<Condition1>** is **FALSE**, then **<Statement4>** is carried out instead.

Be careful of this, otherwise your code will soon become too complex with many **ELSE** statements!

IMPORTANT

If there are many (more than 3) **IF-THEN-ELSE** statements, the PLC code can be difficult to read. A **CASE** statement (chapter 9.2, page 66) can easily replace complex **IF** statements to increase the readability of the code.

It also minimizes the possibility of making errors in complex **IF-THEN-ELSE** statements, when other people correct or add something in the PLC code.

Furthermore, the amount of lines in the PLC code is reduced, when many identical **IF** statements are replaced with a **CASE**. A **reduction** of more than 50 % in the number of PLC code lines is not unusual, when **CASE** is used. See chapter 9.2.3, page 69.

An **ELSIF** statement can be added to check several conditions:

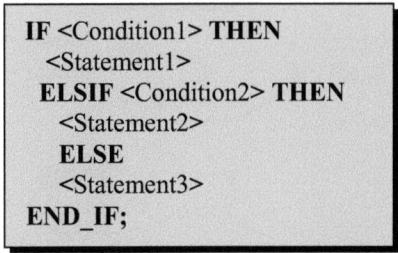

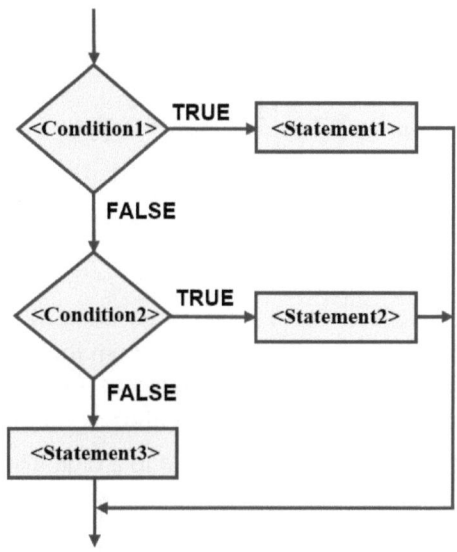

The mode of operation is as follows:

The PLC code in **<Statement1>** will be carried out, if **<Condition1>** is TRUE. If **<Condition1>** is FALSE **<Condition2>** is checked, and if it is **TRUE** the PLC code in **<Statement2>** will be carried out. If **<Condition1>** and **< Condition2>** are FALSE, the PLC code in **<Statement3>** will be carried out.

Flowchart with ELSIF statement

It is recommended to use **CASE** instead of **ELSIF**, because **CASE** creates readable code and flowcharts for **CASE** and **ELSIF** can be the same (see page 66).

9.1.1   EXAMPLE: Motor control with self holding relay

This example shows a motor controlled by a self holding relay (also called latching or keep relay). There is a start switch button with the variable name **S1** which has the data type **BOOL,** and is a **N**ormally **O**pen (NO) contact. In addition, there is a stop switch **S2** with the data type **BOOL**, and this is a **N**ormally **C**losed (NC) contact.

The two manual switches **S1** and **S2** are both supplied with 24V and connected to digital inputs on the PLC, as shown on the diagram below:

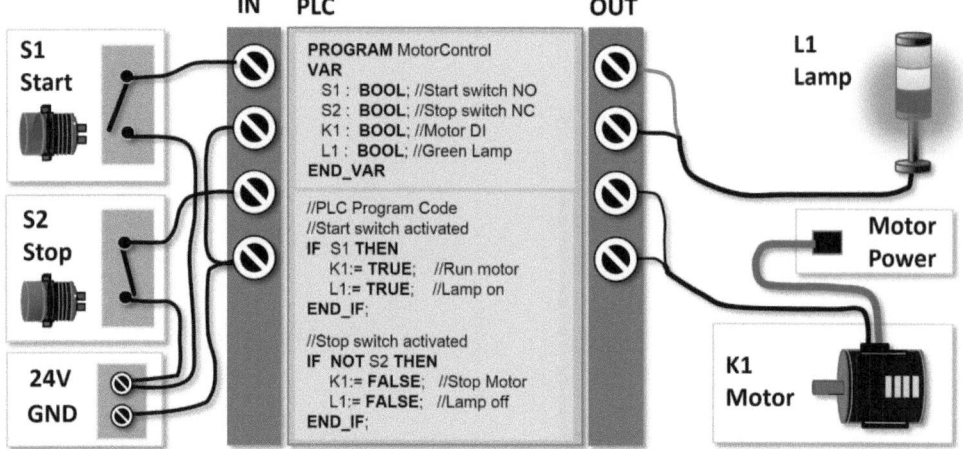

There are two digital outputs from the PLC, one connected to a control lamp **L1**, and the other connected to the motor **K1**. Both **K1** and **L1** have the data type **BOOL** because they are digital outputs.

The output **K1** is connected to a digital input board inside the motor. The motor is configured to run when 24V (**TRUE**) is received on the digital input. This configuration is set by the motor control software tool. The motor stops when 0V (**FALSE**) is received from the input. The motor is supplied with 230V via a separate power supply cable. The **L1** lamp is on when the engine is running.

How the program code works:

When the **S1** switch is activated (button pressed), **K1** is set to **TRUE** and the motor starts. **K1** remains **TRUE**, even though **S1** is no longer activated. If **S2** is activated, **K1** will be set to **FALSE** and the engine will stop. **NOT** (means inverted signal) and is written in front of **S2** in the code, because the **S2** signal is physically shorted in the electrical contact switch, and is therefore normally **TRUE** on the digital input card.

The behavior of the code can be illustrated with a flowchart as shown to the right:

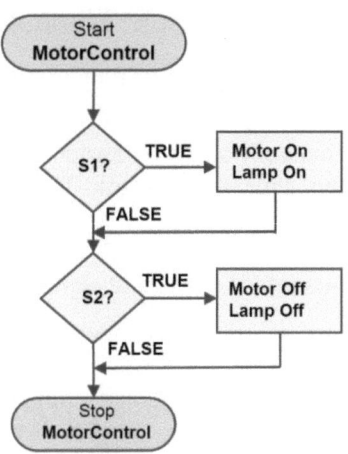

A PLC program executes all the program code in one program scan, after which the variables are set on the output card. This means that if both **S1** and **S2** are activated at the same time, the last value assigned to **K1** and **L1** will be used and set on the output card. Therefore, it is important that the program code for the **S2** stop switch is placed last in the program code.

Program code without IF statement

The program code can be written without using **IF** statements, as shown in the following:

Example **#1** shows how to copy the value of **K1** to **L1**. This means that **L1** is dependent on the output signal of **K1**, which is not good program structure.

```
//#1
K1:= (K1 OR S1) AND S2;
L1:= K1;
```

In example **#2**, there are two almost identical program code lines. This can cause the programmer to forget to change both lines when any further changes are needed.

```
//#2
K1:= (K1 OR S1) AND S2;
L1:= (L1 OR S1) AND S2;
```

The above examples show that only few code lines are needed, but challenges can arise from further changes and extensions to the program.

To obtain a better program structure, an additional variable can be created as shown in example **#3**. Here, an additional temporary variable **T1** is created, and is used to set (assign) the variables **K1** and **L1**.

```
//#3
T1:= (T1 OR S1) AND S2;
K1:= T1;
L1:= T1;
```

Start and stop switch in the same IF statement

Since the program code that stops the motor is placed after the program code that starts the engine, it is not relevant to add **S2** to the **IF** statement, as shown here on the right. Adding **IF** statements would also make the code more complex.

```
//Start switch activated
IF  S1 AND S2 THEN
   K1:= TRUE;
   L1:= TRUE;
END_IF;
```

9.1.2 EXAMPLE: Manually operated tank control

This example shows a manually operated tank control (1) with an inlet valve (2), an outlet valve (5), a motor stirrer (6), and two level sensors (3) & (4).

A manually operated switch **S1** is used to open the inlet valve. When the inlet value is open, liquid fills the tank. The manually operated switch **S2** is used to empty the tank.

Stirring (mixing) can be switched on manually by using switch **S3**.

All variables are digital signals, and therefore declared with the data type **BOOL**.

The diagram delow shows components and associated PLC code:

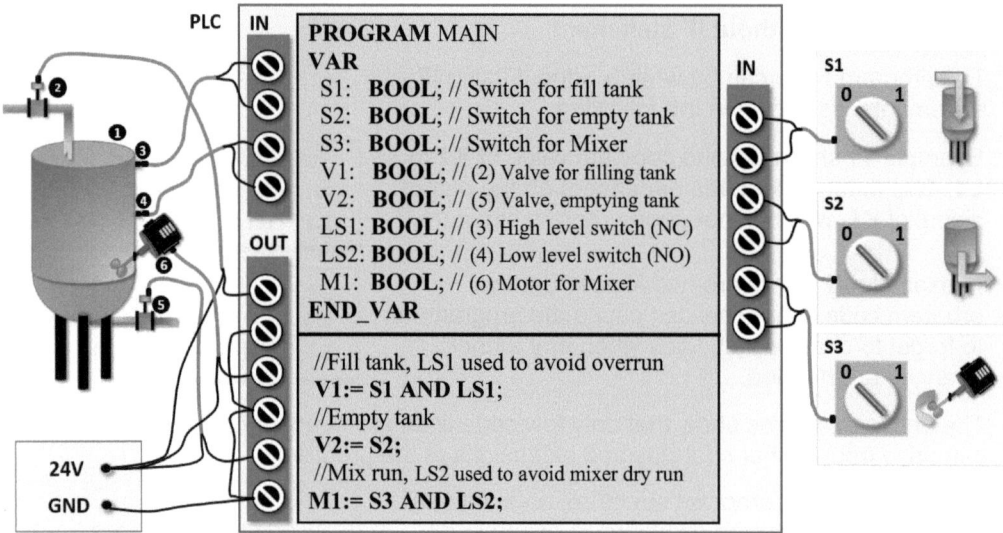

```
PROGRAM MAIN
VAR
    S1:  BOOL; // Switch for fill tank
    S2:  BOOL; // Switch for empty tank
    S3:  BOOL; // Switch for Mixer
    V1:  BOOL; // (2) Valve for filling tank
    V2:  BOOL; // (5) Valve, emptying tank
    LS1: BOOL; // (3) High level switch (NC)
    LS2: BOOL; // (4) Low level switch (NO)
    M1:  BOOL; // (6) Motor for Mixer
END_VAR

//Fill tank, LS1 used to avoid overrun
V1:= S1 AND LS1;
//Empty tank
V2:= S2;
//Mix run, LS2 used to avoid mixer dry run
M1:= S3 AND LS2;
```

The program code has the following built-in safety measures to avoid tank overflow: The inlet valve (2) is connected to the variable **V1**, and the inlet valve cannot be opened when **LS1** is activated. **LS1**, located at the top of tank, is activated by high liquid levels. **LS1** is a Normally Closed (NC) switch, and therefore the signal is **FALSE** if the sensor cable is disconnected, broken or when the liquid level inside the tank is high.

The variable **LS2** is connected to a sensor (4) and if activated, **M1** will not be activated. This ensures the motor stirrer does not run if there is no or very low liquid levels inside the tank, also known as a dry run. If there is no liquid in the tank the motor stirrer may get hot and malfunction. **LS2** is conneced to a Normally Open (NO) switch and will therefore become **TRUE** when liquid reaches the sensor (4).

9.1.3 EXAMPLE: IF-THEN-ELSE open and close valve

The following PLC code example checks the alarm from a pump and the pressure in relation to a set-point:

```
IF (PumpAlarm = TRUE) AND (PumpPressure > PumpSetPoint) THEN

    ValveOpen := TRUE;   //Open valve
ELSE
    ValveOpen := FALSE; //Close valve
END_IF;
```

If the whole condition, marked by white frame in the **IF** statement, is fulfilled (**TRUE**), the valve **ValveOpen** is opened, otherwise the valve is closed.
PumpAlarm can be a digital input signal with the data type **BOOL**.
PumpPressure is a variable with the data type **REAL,** and can be an analog input that receives a signal from a pressure sensor.
PumpSetpoint (the desired pressure setting that the pump should control) is a **REAL** data type, and can be a value that the user can adjust via a user control panel (HMI).

The above PLC code example can be rewritten to:

```
ValveOpen := FALSE;   //#1 Note

IF (PumpAlarm = TRUE) AND (PumpPressure > PumpSetPoint) THEN
    ValveOpen := TRUE;   //#2 Note
END_IF;
```

This means a code line less and is simpler to read for some programmers.

NOTICE:
Values are only moved to the output modules, when *all* PLC code has been executed (a program scan), therefore the connected valve **ValveOpen** will not close straight away (see **#1**) and open (see **#2**) right after.

To make the PLC code simpler, it can be rewritten as follows:

```
ValveOpen := PumpAlarm AND (PumpPressure > PumpSetPoint);
```

The variable **ValveOpen** is set to **TRUE** or **FALSE**, without using an **IF** statement!

9.1.4 EXAMPLE: Robot control for packing items

This example is of a small robotic plant where a robot packs three items in a box. Afterwards, the box is checked by a vision camera:

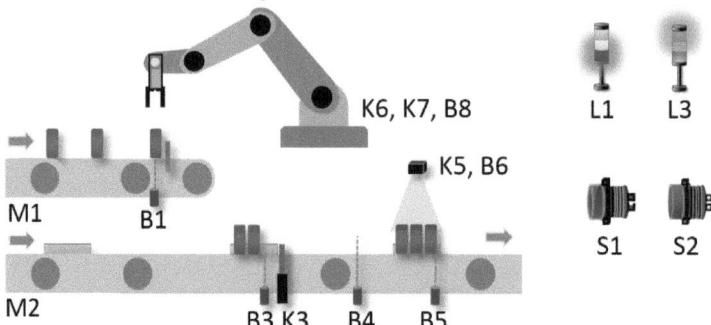

The robot has its own control system (robot controller) which is controlled from the PLC by two digital signals (K6, K7). When the robot has finished moving the items to the box, the PLC receives a confirmation signal (B8) from the robot controller. Finally, the box is checked by a vision camera. The conveyor belts are controlled by the PLC.

Components and mode of operation:

Name	I/O	Component	Mode of operation
S1	DI	Start switch	When activated, the plant starts: Conveyor belts are running and the robot is moving items.
S2	DI	Stop switch	(NC – Normally Closed contact). Stops plant running.
B1	DI	Sensor	Signals when an item can be removed by robot.
B3, B4, B5	DI	Sensor	Signals when a box breaks the sensor light beam. **B3** starts robot program. **B5** launches vision camera. **B4** activates **K3**.
B6	DI	Vision	Signals if the camera confirms that the box is ok.
B8	DI	Robot	Signals when Robot has moved three items.
K3	DO	Air Cylinder	On signal: The cylinder has moved up and the box is stationary, so the robot can place items in the box. Controlled by **B4** and **B8**
K5	DO	Vision camera	On signal: Take a picture of the box and compare with previous pictures of boxes which are ok.
K6	DO	Robot	Robot start signal. A new box is ready to be filled. Robot stops when no signal is received.
K7	DO	Robot	Moves an item to the box.
L1	DO	Green light	Latest box meets requirements. (Box ok)
L3	DO	Red light	Latest box does not meet requirements (Box not ok)
M1, M2	DO	Conveyor belt	Powered by a motor with a frequency converter. Speed is set directly on the frequency converter.

All variables are digital signales with the data type **BOOL**.

Below find program code and flowcharts:

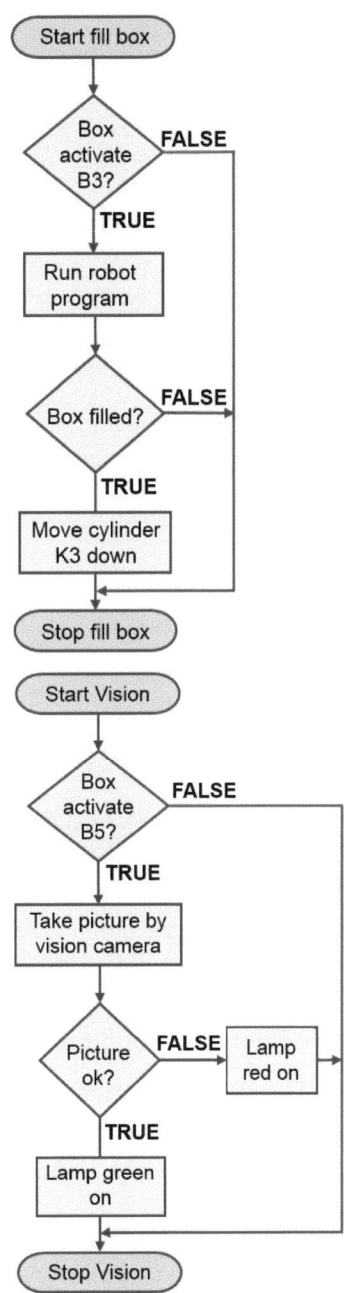

```
//////////////////////////////////////////////////
//Program code for box filling by Robot
//////////////////////////////////////////////////
//Note: All variables are BOOL data types

//Start button pressed
 IF S1 THEN
   M1:= TRUE;  //Start Conveyor belt 1
   M2:= TRUE;  //Start Conveyor belt 2
   K3:= TRUE;  //Cylinder up, box filling stopped
 END_IF;

//Start button pressed (NC contact)
IF NOT S2 THEN
   M1:= FALSE;  //Stop Conveyor belt 1
   M2:= FALSE;  //Stop Conveyor belt 2
   K6:= FALSE;  //Stop robot program
END_IF;

//Start box filling program (refer to flowchart)
IF B3 THEN
   K6:= TRUE;  //Run robot program

   //Robot is running moving items to box
   K7:= B1 AND K6;

   // Robot program done
   IF B8 THEN
     K6:= FALSE;  //Stop robot program
     K3:= FALSE;  //Cylinder down. Let the box pass
   END_IF;

   // Cylinder up. Ready for next box
   IF B4 THEN
     K3:= TRUE;
     K5:= FALSE;  //Ready for next vision check
   END_IF;
END_IF; //End box filling

//Check box by camera (refer to flowchart)
//Light is on until next check
IF B5 THEN
   K5:= TRUE;  //Box activates camera to take picture
   L1:= B6;  //Green light
   L3:= NOT B6;  //Red light
END_IF;
```

9.2 CASE statement

CASE is a statement used when different events or actions are to be carried out based on only one variable. Use **CASE** when **IF** statements become too complex.

CASE is good for sequence control (also called a sequencer or finite-state machine (FSM)) and is often applied (used), when e.g. a machine can be set in different operational modes (e.g. STOP, STARTING, RUN, STOPPING) or applied in a process in a dairy (e.g. NONE, CREAM, SKIM_MILK, WHOLE_MILK, WATER_FLUSH).

A **CASE** statement has the following syntax:

```
CASE <Condition> OF
   <Condition1> : <Statement1>;
   <Condition2> : <Statement2>;
   <Condition3> : <Statement3>;
   ...
ELSE
   <Statement>
END_CASE;
```

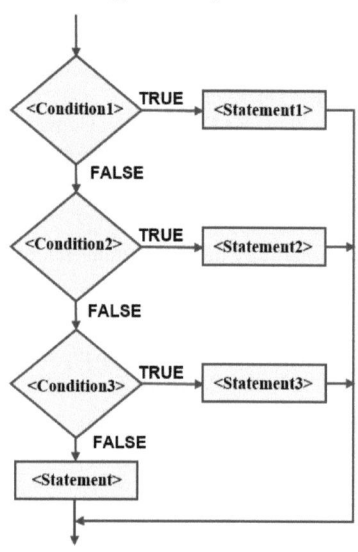

Mode of operation:

The variable to be checked is the **<Condition>** and this must be an integer data type (**INT**, **DINT**, **WORD**).

The different values **<Condition>** can take, are written in the sections **<Condition1>**, **<Condition2>** and **<Condition3>** followed by colon ":"

The action to be executed is written in **<StatementX>**, here **X** is 1, 2 or 3 which can be PLC code. If the PLC code is longer than 4 to 6 lines, a function or a program module should be created with readable program code.

If **<Condition>** = **<Condition2>** the code in **<Statement2>** will be executed.

It is not required to have PLC code in **<StatementX>**. The sections can be empty.

The three dots (...) indicate that the number of sections is unlimited, but must be at least one **<Condition3>** : **<Statement3>** line in each **CASE** statement.

The **ELSE** section is optional. It is, however, recommend that some PLC code is written in this section, like e.g. an alarm/error message, so that the programmer is aware that a program call is performed in the **ELSE** section.

9.2.1   EXAMPLE: CASE – Setting the motor speed

Here is an example of how to use a **CASE** statement, where the speed of a motor is adjusted on a selector switch named **MotorSwitch**. The switch can be activated in steps from 1 to 6, which can be 6 voltage levels. **MotorSwitch** is an **INT** data type.

```
MotorFan:= FALSE;  //Turn off the motor cooling

CASE MotorSwitch OF
  1, 2 : MotorSpeed := 25;  //Two values, separated by comma
  3    : MotorSpeed := 35;  //One value in CASE
  4..6 : MotorSpeed := 50;  //Interval CASE: start no. .. end no
         MotorFan:= TRUE; //Turn on the motor cooling
ELSE
  MotorSpeed := FALSE;  //Use as default
END_CASE;
```

Selector switch/ Rotary switch

Explanation of example:

If **MotorSwitch** is 1 or 2, the **MotorSpeed** will be 25. If **MotorSwitch** is 3, the **Motor Speed** will be 35. If **MotorSwitch** is 4, 5 or 6, the **MotorSpeed** will be 50.

If no **CASE** condition is fulfilled, i.e. **MotorSwitch** is not 1 to 6, the **MotorSpeed** will be set to zero (the motor is not running).

The **MotorFan** variable is always set to **FALSE** (ventilation turned off) before the **CASE** code begins. Only when **MotorSwitch** is 4 to 6 the **MotorFan** is set to **TRUE**, where the cooling has to run. When **MotorFan** is set to **FALSE** before the **CASE** code, it eliminates the need to write **MotorFan:= FALSE** in all other **CASE** sections.

The **ELSE** statement ensures that **MotorSpeed** will be zero, if **MotorSwitch** has a value which the **CASE** does not recognize, - this ensures better quality PLC code and provides instructions in the event of unknown values of **MotorSwitch**.

Because there is a risk that the PLC programmer 'forgets' to change all the values in the PLC code, if values need changing, it is recommended to replace the values 1, 2, 3, 4 and 6 with variables created as **CONSTANT**, as the values is used more than once in the same PLC code. Read more about **CONSTANT** in chapter 6.2 page 38.

9.2.2 EXAMPLE: CASE – For executing programs

In this chapter **CASE** is used to execute different program modules. See more about splitting code into program modules in chapter 10, page 78.

The example is shown without and with the use of **CONSTANT**.

The value of **ProgramSelect** determines which program module will be executed:

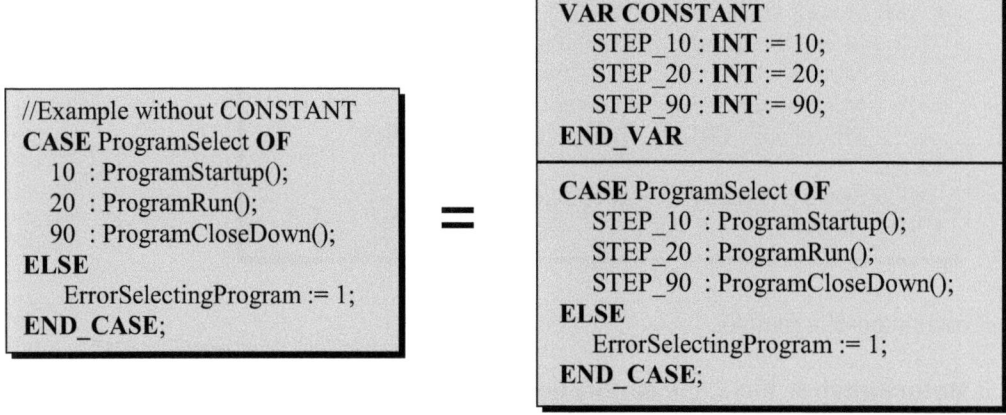

```
//Example without CONSTANT
CASE ProgramSelect OF
    10 : ProgramStartup();
    20 : ProgramRun();
    90 : ProgramCloseDown();
ELSE
    ErrorSelectingProgram := 1;
END_CASE;
```

=

```
PROGRAM Main
VAR CONSTANT
    STEP_10 : INT := 10;
    STEP_20 : INT := 20;
    STEP_90 : INT := 90;
END_VAR

CASE ProgramSelect OF
    STEP_10 : ProgramStartup();
    STEP_20 : ProgramRun();
    STEP_90 : ProgramCloseDown();
ELSE
    ErrorSelectingProgram := 1;
END_CASE;
```

Mode of operation:

If **ProgramSelect** is 10, then the program module **ProgramStartUp** is executed. Inside the **ProgramStartUp**, the variable **ProgramSelect** is changed to 20, so that **ProgramRun** is executed in the next program-scan instead of **ProgramStartUp**.

If the condition variable **ProgramSelect** is set to a value not implemented in the **CASE** statement, the variable **ErrorSelectingProgram** is set to 1, to inform the programmer that a program module has not been selected.

The fixed values **ProgramSelect** can take, are: 10, 20 or 90. A jump between the values (11, 12, 13 .. to 19) is intentionally created to give space for future extensions. The values can be created as **CONSTANT** or **ENUM** so the values are easy to find in the PLC code, and can be changed in one place only.

IMPORTANT   A good software structure uses **CASE** statements to execute different program modules. **CASE** statements gives a better overview than many **IF**-statements, especially **ELSE-IF** statements.

9.2.3   EXAMPLE: CASE – Recognizing numbers

The example below shows how the **CASE** statement can be used to recognize numbers. The numbers could be a password, which is used to give a user access to the user control panel (HMI). There are often various levels of access such as:

>Operator password
>Administrator password
>SuperUser password

In the following example a variable **PassOK** is used to check whether **PassSelect** contains a valid password or not.

In the first program line, the variable **PassOk (BOOL)** is set to **FALSE.**

If the variable **PassSelect** contains one of the values 1747, 3309, 5607, 1234 or 1027 the variable **PassOK** is set to **TRUE** as shown below:

```
PassOk := FALSE; //No valid password number

CASE PassSelect OF
   1747, 3309, 5607, 1234, 1027: PassOk := TRUE //Valid password number
END_CASE;
```

The example shows that **CASE** is a simpler solution than applying many **IF** statements, because the above solution will require five **IF** statements (15 lines of PLC codes) or a very long **IF** statement, as shown in the example below:

```
PassOK := FALSE;  //Default, if not found in the IF line below

IF PassSelect = 1747 OR PassSelect = 3309 OR PassSelect = 5607 OR
      PassSelect = 1234 OR PassSelect = 1027 THEN
   PassOK := TRUE; // Valid password number
END_IF;
```

As can be seen above, long lines of code can be written, but this will make the PLC code difficult to read. It is recommended that the PLC code lines are no longer than the screen width of the PLC compiler programming tool.

9.3   Iteration statement, LOOPS

Loops are used to repeat PLC code a number of times. Loops are often used when all values in an **ARRAY** must be set to a specific value, or a maximum or minimum value must be found in an **ARRAY**.

It is important to ensure that DEAD LOCK does not occur in the PLC. This is a situation where the CPU uses all its power to work on the loop, and is a common programming error. To ensure that the loop ends, it must end after a specified amount of time or a specified number of executions.

The next chapter shows different methods of implementing loops.

9.4   FOR-DO Statement

This type of loop is the most frequently used type. A FOR-DO is always executed a certain number of times. This is determined by a start value and an end value.

Syntax is as shown below:

```
FOR <ValueStart> TO <ValueEnd> DO
  <Statement>
END_FOR;
```

Where:

<ValueStart> =	A counter variable (**INT** or **WORD**) set to a start value.
<ValueEnd> =	Execute the **<Statement>** programming code until the counter variable reaches this value
<Statement> =	Containing the PLC code to be executed. It can be one or more PLC code lines. It is recommended that lines between **FOR** and **END_FOR** is indented with 2 X SPACE. This makes the code more readable.

NOTICE   It is not allowed to change the counter variable in the **<Statement>** section – it interrupts the program execution!

A FOR-DO statement always adds 1 to the count value per execution. If there is a need to add more than 1, **BY** is added. This is, however, not used very often.

If the loop needs to step backwards (**ValueStart > ValueEnd)** use **BY** -1

Syntax for using **BY**:

```
FOR <ValueStart> TO <ValueEnd> BY <ValueStep> DO
   <Statement>
END_FOR;
```

Where:

> **<ValueStep> =**   **OPTION.** The step value if different from 1
> If the loop needs to step backwards use BY -1

NOTICE: The smaller PLC types with less calculation capacity, cannot handle large FOR-DO statements because this can create long scan-times. When using the small PLC types, it is recommended to reduce the FOR-DO statements, or to split up the FOR-DO statements into smaller FOR-DO statements, and place these in different program modules and execute them with different scan-times.

IMPORTANT
A typical error when working with FOR-DO statements and **ARRAY** is that the first or last position in the **ARRAY** are not handled. Another typical error is when the loop runs longer than the size of the **ARRAY,** which can result in unstable PLC code.

Variable names such as i, j, n or m are often used as counter variables.

If there is a need of exiting the FOR-DO statement before all loops are carried out, the **EXIT** command can be used. This is used if the task is to find a value inside the **ARRAY** and when the value is found, the program can exit the loop.

Syntax for using the **EXIT** command:

```
FOR <ValueStart> TO <ValueEnd> DO
   <Statement>
   IF <Condition> THEN //#1, Exit now?
      EXIT;          //Exit the loop
   END_IF;
END_FOR;
```

As shown above, **IF** must be added (**#1**) within the FOR-DO statement and if the **<Condition>** is **TRUE**, the **EXIT** will be carried out and the loop immediately ends.

9.4.1   EXAMPLE: FOR – A loop running 4 times

In this chapter the example shows an **ARRAY** with 4 elements of the data type **INT**. Each of the elements is set to 7 by using a FOR-loop:

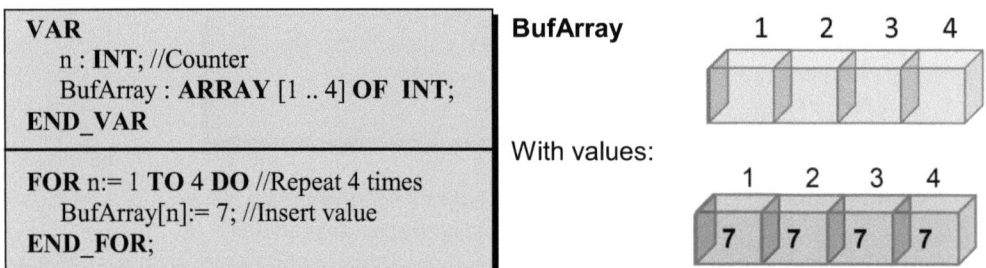

```
VAR
   n : INT; //Counter
   BufArray : ARRAY [1 .. 4] OF  INT;
END_VAR

FOR n:= 1 TO 4 DO //Repeat 4 times
   BufArray[n]:= 7; //Insert value
END_FOR;
```

BufArray

With values:

Number 1 and 4 occur twice in the example, when **ARRAY** is created and in the FOR-loop. These numbers must, therefore, be created as a **CONSTANT**, as a typical error is the programmer forgetting to change the value in both places in the code.

The counter variable **n**, used by the FOR-loop, counts (adds) 1 each time the loop executes. In each execution, the value 7 is inserted on the position in **BufArray**, in the relevant variable **n**.

The example above can be rewritten to the following four lines of code:

```
VAR
   BufArray : ARRAY [1 .. 4] OF  INT;
END_VAR

BufArray[1] := 7;
BufArray[2] := 7;
BufArray[3] := 7;
BufArray[4] := 7;
```

The FOR-loop replaces four lines of the PLC code!

Single values can be inserted directly into **BufArray** as follows:

```
BufArray[2] := 23;
BufArray[3] := 12;
```

9.4.2 EXAMPLE: FOR – LOOP and 3D ARRAY

The example in this chapter shows how all elements in a 3-dimensional **ARRAY** named **Array3D** are set to 1. This method can be used right after the program starts, or if all positions in an **ARRAY** must be assigned (set to) a certain value.
In the variable section three variables are created (declared): x, y and z, which are used to index the **ARRAY**.

To define the size of the **Array3D**, three **CONSTANT** variables are created: **X_MAX**, **Y_MAX** and **Z_MAX**, so it is easy and safe to change the size of the **ARRAY** later. **ARRAY** might have another size during test and the implementing of the PLC code, and by using **CONSTANT** all positions are changed.

```
PROGRAM MAIN
VAR CONSTANT
    X_MAX : INT := 10;
    Y_MAX : INT := 20;
    Z_MAX : INT := 30;
  END_VAR
  VAR
    x, y, z : INT; //Index to the 3D Array
    Array3D : ARRAY [1 .. X_MAX, 1 .. Y_MAX, 1 .. Z_MAX] OF INT ;
  END_VAR

  FOR x:= 1 TO X_MAX DO
    FOR y:= 1 TO Y_MAX DO
      FOR z:= 1 TO Z_MAX DO
        Array3D[x, y, z] := 1;     //Set current position to 1
      END_FOR;
    END_FOR;
  END_FOR;
```

A 3D **ARRAY** can e.g. be used for: placing packages on a pallet in a production line, a large warehouse, a logistics terminal or a big parking garage.

In the above example, 10 x 20 x 30 = 6000 elements with **INT** variables are created. It can result in execution problems in smaller PLC types, when a loop with 6000 elements are executed. If this occurs, a number of 2D **ARRAY** can be created instead, as a 3D **ARRAY** can always be rewritten to a number of 2D **ARRAY**.

9.4.3   EXAMPLE: Calculation of the average value

The following example shows how a FOR-DO loop can be used to calculate an average value of a range of values saved inside an **ARRAY**. It is assumed that the averaged values are already saved in **BufArray**:

```
PROGRAM Average
VAR CONSTANT
  BufArrayMin : INT := 0;
  BufArrayMax :INT := 9; //Must be higher than BufArrayMin
END_VAR
VAR
  i                    : INT;    //Counter variable in FOR LOOP
  BufArray             : ARRAY [BufArrayMin .. BufArrayMax] OF REAL;
  BufArraySum          : REAL;   //Calculator for the value sum
  BufArrayAverage      : REAL;   //Average value of the BufArray
END_VAR

BufArraySum := 0; //Reset calculator #1)

//Sum all values from the buffer into BufTempVar #2)
FOR i := BufArrayMin TO BufArrayMax DO
  BufArraySum := BufArraySum + BufArray[i];
END_FOR;

//Calculate average
BufArrayAverage := BufArraySum /( BufArrayMax – BufArrayMin + 1 );
```

Overview of **ARRAY** with 10 positions (elements):

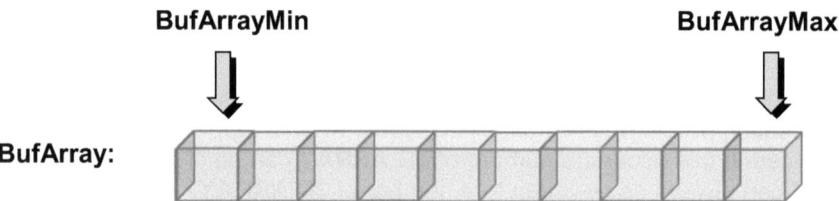

BufArray:

Explanation to the **Average** example program:

Constants (VAR CONSTANT)
Two constants, **BufArrayMin** and **BufArrayMax**, are created (declared), because the constants are used three times in the PLC code, and when changing the length of the **BufArray** the constants ensure all values are changed.

Naming
The first part of constants and array shares the same name, **BufArray**, which indicates that they belong together.

Mode of operation:

BufArraySum is a variable to contain the sum of the all values.

First, the variable **BufArraySum** is initialized to the value zero (0) to ensure that the content is zero the first time it runs. #1)

The next step adds up all values in the **BufArray** by using a FOR-DO loop. **BufArrayMin** is at the first position (the start of **BufArray**) and ends with **BufArrayMax** (the end of **BufArray**). Notice that the number of times the FOR-DO loop executes is **BufArrayMax – BufArrayMin + 1** as the first and the last execution in the loop are both included. #2)

When the loop is finished the variable **BufArraySum** now contains all values, added together. To find the average value, the total amount is divided by the number of the times the FOR-DO loop has executed. The result can be found in the **BufArrayAverage** variable.

It is important to make sure that the result of the calculation **BufArrayMax – BufArrayMin + 1** time is not zero, because a PLC cannot handle a division by zero.

The calculation of the average value in a PLC is often used to filter input signals from analogue sensors. When filtering signals, 'noise' might be removed from the recorded values. The disadvantage of a FOR-DO loop for this purpose is that array requires a lot of memory, takes up CPU time and the calculation of an average is based on all values in the array. Therefore, it can be advantageous to use a digital filter. See more in chapter 13.4, page 134.

9.4.4   EXAMPLE: Find the lowest value in an array of numbers

This example uses a **FOR** loop to find the minimum value in an array of numbers.

The task is to find the lowest number in the row shown to the right. Here the lowestminimum number is 12 which can be found in position 3:

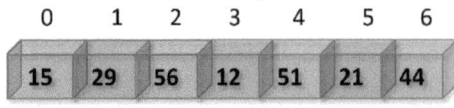

To find the minimum value (lowest number): Select the first value in the row. This value is 15 (where the variable **i** is 0). Now the **FOR** loop will compare this value to the next in the row. If the next value is lower, the value is saved and used as the new minimum value. An **IF** statement is used to check if the value in the row is lower than the saved one. The **FOR** loop continues (loops through) until all the numbers in the row have been compared.

LOOP Tabel	
I	MinValue
0	15
1	15
2	15
3	12
4	12
5	12
6	12

Below find the program code and flow chart:

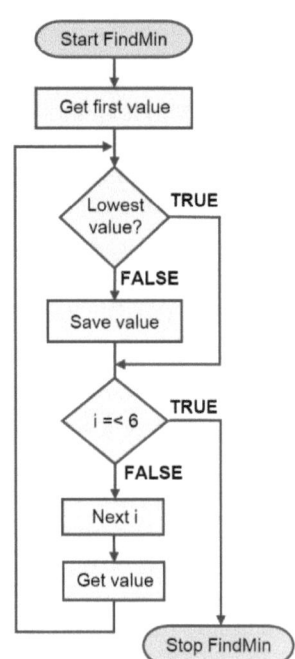

```
PROGRAM FindMin
VAR
  Ar: ARRAY[0 .. 6] OF INT := [15, 29, 56, 12, 51, 21, 44];
  MinValue: INT;        //The minimum value found
  i:        INT;        //Counter for the LOOP
  MinIndex: INT := -1;  //Index where min. value was found
END_VAR

//Save the first value to have a value to compare to
MinValue:= Ar[0];

//For all values in the ARRAY.  Note: Start from index 1,
// (position 1) because the first value was set as the start value
FOR i:= 1 TO 6 DO
  //Is the next value in the row lower than the saved value?
  IF Ar[i] < MinValue THEN
    MinValue:= Ar[i]; //Save value
    MinIndex:= i; //Save index
  END_IF;
END_FOR;
```

The minimum value in the array is 12, and is saved in the **MinValue** variable. It was found in position (index number) 3, which is saved in the variable **MinIndex**.

9.4.5   EXAMPLE: Sorting numbers inside an ARRAY

The following section covers an example of how to sort numbers in an **ARRAY**. Sorting is carried out by copying the numbers to a new **ARRAY**, organised from lowest to highest number

This example uses the program code from the previous page where the lowest value was found in an **ARRAY**. The program code is now executed 7 times because there are 7 numbers in the **ARRAY**. When the lowest number is found, the number 999 replaces the number in its position to make sure the number does not appear in the search again.

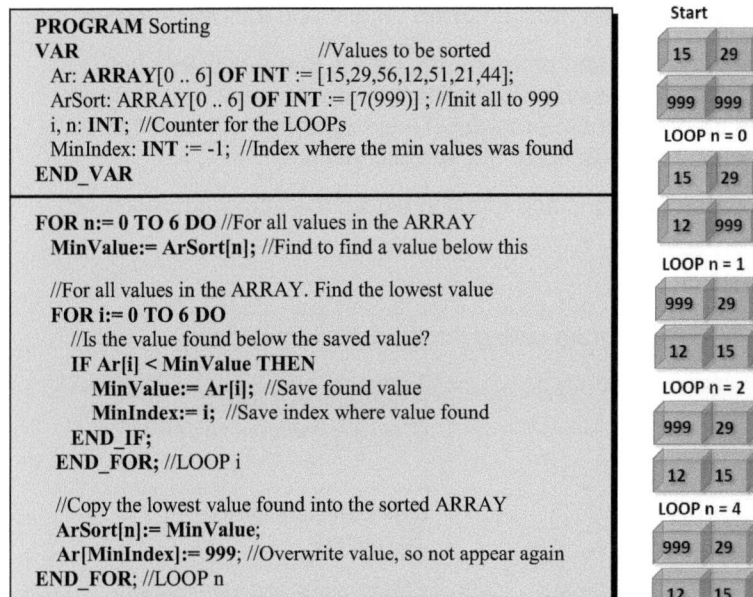

```
PROGRAM Sorting
VAR                              //Values to be sorted
  Ar: ARRAY[0 .. 6] OF INT := [15,29,56,12,51,21,44];
  ArSort: ARRAY[0 .. 6] OF INT := [7(999)] ; //Init all to 999
  i, n: INT;  //Counter for the LOOPs
  MinIndex: INT := -1; //Index where the min values was found
END_VAR

FOR n:= 0 TO 6 DO //For all values in the ARRAY
  MinValue:= ArSort[n]; //Find to find a value below this

  //For all values in the ARRAY. Find the lowest value
  FOR i:= 0 TO 6 DO
    //Is the value found below the saved value?
    IF Ar[i] < MinValue THEN
      MinValue:= Ar[i];  //Save found value
      MinIndex:= i;  //Save index where value found
    END_IF;
  END_FOR; //LOOP i

  //Copy the lowest value found into the sorted ARRAY
  ArSort[n]:= MinValue;
  Ar[MinIndex]:= 999; //Overwrite value, so not appear again
END_FOR; //LOOP n
```

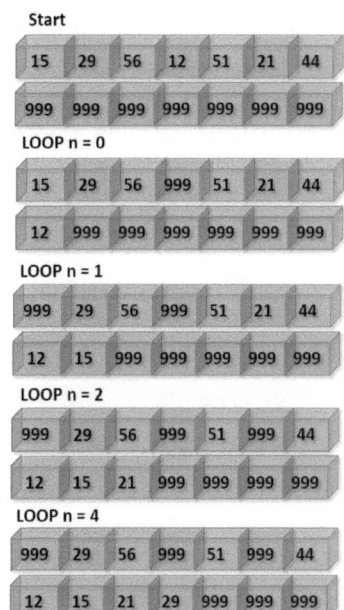

There are many methods of sorting numbers and the example shown here is just one of them. The disadvantage of this example is that the original values are overwritten by the value 999. To mitigate this you can retain the original values, by copying them to another **ARRAY** before sorting.

IMPORTANT: The **ARRAY** size, which is 6, must be declared as a **CONSTANT**.

10 Splitting up the PLC program

A PLC program must be split up into many small program pieces in order to have a good and clear program structure. The small program pieces are called program modules, functions and function blocks, and they each contain a small piece of PLC code and is a building block, to be used or reused whenever needed.

To obtain a good structure, it is a good rule of thumb to only have only 20 - 25 lines of PLC code in each program module, function or function block.

When the PLC program is split up into pieces, the execution order can easily be changed, and program modules and functions can easily be made inactive during fault-finding (debugging) (done by putting the // characters before the name). Furthermore, it is much easier to work with many small pieces of PLC code, rather than one large giant program, and it is easier to move and fix small pieces of code.

The program modules and functions must be given unique and indicative names.

The difference between functions and program modules is that functions often perform calculations or data processing on individual components, whilst program modules is the splitting up of the entire program. The program modules use relevant functions and function blocks to solve the specific tasks.

It is typically easier to reuse functions and function blocks than program modules.

10.1 Programmodules

Below diagram shows a main program calling three program modules:

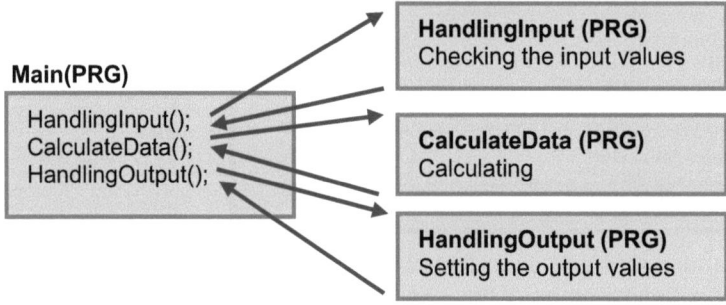

The main program **Main** is executed once in each program-scan. In **Main,** the program module **HandlingInput** is listed first and therefore the code inside this program module is executed first. Then the code inside the **CalculateData** program module is executed. Finally, the **HandlingOutput** program module is executed.

When one program-scan ends, it will be repeated again - the **Main** program is called again. If the program-scan time is 50 [ms], the **Main** program is executed every 50 [ms]. It is important that the total time of execution for all four program modules is less than 50 [ms]. If the program modules contain large arrays or many calculations, all program modules might not be executed within the scan-time. To solve this issue, the program-scan time has to be increased or the programs checked to see whether code can be reduced or redesigned. Remember that the length of program scan-time can vary depending on the number of **IF**-statements or **CASE**-statements in the code. Therefore, the longest possible scan-time must be accounted for in the code, and add some extra time just in case.

Normally the PLC will inform the programmer if the scan time needs to be increased.

The program modules can be configured to have different scan times. This is useful because not all program modules take the same length of time to execute:

Main (PRG)
Scan time: 50 [ms]
Main program

Time1Secound(PRG)
Scan time: 1000 [ms]
Update timer

AvgCal(PRG)
Scan time: 10000 [ms]
Calculate average values

TemperatureCal(PRG)
Scan time: 2000 [ms]
Calculate temperatures

By splitting the program into program modules, these advantages are achieved:

- Better and clear program structure
- Possibility to configure individual scan time
- Easy to change program execution order
- Program modules can be set to inactive when troubleshooting

There are many ways of splitting code into program modules. Below find inspiration:

- Sensors mounted on one side of the machine
- Digital input from electrical switches and circuit breakers
- All motors used for ventilation
- Value preparation to/from the HMI (user control panel)
- Alarm surveillance / alarm supervision
- Data communication to other PLCs

10.2   Functions

Functions are important building blocks in a PLC program. A function contains a limited number of code lines to be used ('called' and executed) again and again.

Functions are also called: procedures, sub-routines or add-ons.

A function's 'call' to **MyFunction** is carried out as follows:

```
MyFunction ();
```

The above function 'call' does not take parameters, as the brackets are empty. Functions can take one or more input values (parameters), on which the function will work on or use to perform calculations.
When the function 'call' has ended, one or more values (parameters) is returned by the function. The returned value can then be used by the rest of the program.

Below shown a function 'call' which takes two parameters (input values), 12 and 3:

```
MyFunction1 (12, 3);
```

Calculations can also be made before a function 'call'. In the below example two numbers (3+7) are added up just before the function is 'called':

```
MyFunction2 (3 + 7);
```

The calculation is carried out before the function 'call' and the input value that enters the function is therefore 10.

If the function is to return values when the function has ended (a result of one or more calculations), this can only be carried out by using variables, as the function has a 'shelf' in the memory to deliver the value to. When a function is 'called' with an input variable, the function will collect the value from the variable 'shelf' in the memory and deliver a copy of the variable to the function.

The advantage of using functions is that the PLC code can be reused. PLC code re-use reduces the size of the program, creates fewer syntax faults and is easier to work with for other programmers.

Below is shown a program 'call' to a function with variables:

> **MyFunctionInOut (Var1:= ValueIn, Var2=>ValueOut, Var3:= ValueInOut);**

The three variables **Var1**, **Var2**, **Var3** are created (declared) inside the function with the following variable scope:

Variabel	Scope	Assignment
Var1	IN	:=
Var2	OUT	=>
Var3	IN_OUT	:=

Block diagram of the function:

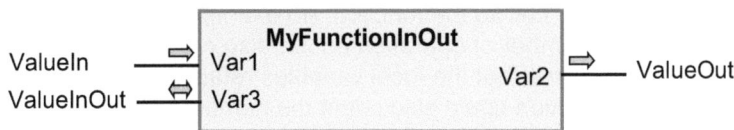

ValueIn is a value going into the function and is written as follows:

> **Var1:= ValueIn;**

The value which is going out of the function must be delivered in the variable **ValueOut** as follows:

> **Var2=>ValueOut;**

The variable which is both going in and out of the function delivers the address pointing (a link) to the 'shelf' in the memory and is written as follows:

> **Var3:= ValueInOut;**

Notice how the assignment signs "**=>**" and "**:=**" are used at function 'calls'.

How to make a function 'call' an ARRAY:

A program call to index no. 4 in an **ARRAY** in a function is carried out as follows:

> **MyFunction [4] (ValueIn);**

10.3 Function (FC) and Function Block (FB)

There are two function types in a PLC:

Function (FC)
Function block (FB)

Function (FC) PLC code excludes static data, which means that all local variables lose their value when the function ends. The variables are initialized again the next time the function is 'called'. The function typically carries out a mathematical calculation and returns the calculated value.

Function block (FB) PLC code which includes static data. The local variables retain their values between each 'call' to the function. An example could be a function used as an hour counter (number of operation hours, also called TACHO HOURS) on a motor which requires that the local variables retain their values once the function has ended. The function could also count the number of motor starts per hour or time until the next motor service.

Syntax of a function (FC):

```
FUNCTION <Name> : <RetDataType>
  VAR_INPUT
    <Variables>
  END_VAR
  VAR_OUTPUT
    <Variables>
  END_VAR
  VAR_IN_OUT
    <Variables>
  END_VAR
  VAR
    <Variables> //Local variables
  END_VAR
    <Implementation> //Write code here
    <Name> := 123;    //Set return value
END_FUNCTION
```

Syntax of a function block (FB):

```
FUNCTION_BLOCK <Name>
  VAR_INPUT
    <Variables>
  END_VAR
  VAR_OUTPUT
    <Variables>
  END_VAR
  VAR_IN_OUT
    <Variables>
  END_VAR
  VAR
    <Variables> //Local variables
  END_VAR
    <Implementation> //Write Code here
END_FUNCTION
```

The implementation of a function starts with the keyword: **FUNCTION** and a function block starts with the keyword: **FUNCTION_BLOCK**. Then comes the name of the function written in the **<Name>** field. It must be an indicative name (see chapter 6, page 31) related to what task the function performs. The return data type is written in the **<RetDataType>** field, as the name of the function works as the return value.

Notice that the **<RetDataType>** field cannot be used in **FUNCTION_BLOCK**.

The sections with **VAR_INPUT**, **VAR_OUT** and **VAR_IN_OUT** must contain variables which go in and out from the function. When **VAR_INPUT** is used, the function copies the variable and works on it inside the function – without overwriting the original variable value. **VAR_IN_OUT** must be used carefully as the address (a link) of the variable is delivered to the function which performs calculations directly on it inside the function changing its value permanently.

If a function needs to work with **STRUCT** or **ARRAY** the **VAR_IN_OUT** must be used.

The order in which variables are listed inside the function indicates the order in which the variables will appear when 'calling' the function.

The section **VAR** contains the local variables, only to be used internally in the function. When a function 'call' is carried out, the local variables are created every time the function is 'called' and deleted again when the function has ended. Remember that variables must be initialized (be set to a start value, e.g. 0) to ensure the correct value of the variables, when the function is 'called'.

If the function needs to save local variables at each 'call', either a **FUNCTION_BLOCK** or **VAR_IN_OUT** must be used, so that the function works on variables created outside the function.

The PLC code, the function is to execute, is written in the **<Implementation>** section.

When a **FUNCTION** is used, the return parameter must be set *BEFORE* the function ends. It is set by using the name of the function which will be assigned the return value. In the above syntax, the return parameter is set to 123. Only one return parameter can be set in this way. If more return parameters are needed, use the **VAR_OUT** and/or **VAR_IN_OUT**.

10.4 Design guide for implementation of a function

This chapter contains tips and a guideline to implement a function.

The aim when developing functions is to achieve fewer program errors and create reusable program code resulting in a more structured program.

To create reusable code, a function:

1	May not use variables directly from a program module or global variables. The function must not have direct access to specific I/O modules. Variables must be accessed by using **VAR_IN**, **VAR_OUT** and **VAR_IN_OUT**.
2	Must have a general name which indicates what the function does. It should preferably not be the same as the specific PLC type, company name, your name or specific sensor numbers and sensor names. Create an independent, but indicative name.
3	Must be created so it can be used by another PLC type and programming language.
4	Should be able to be called from both a program module and another function.
5	Should only have a maximum of 8 IN-OUT parameters (variables), as it can be difficult to maintain a good overview of a large number of variables. If multiple variables are needed, a **STRUCT** can be created to group them. If **STRUCT** is used, the variables must be created in the **VAR_IN_OUT** section. Typically, a function has 1 to 4 input parameters and 1 output parameter. A description of the **STRUCT** must be included in the documentation for the function.
6	Must check if all variables entered into it are valid. A function must not become unstable or produce a Run Time Error (RTE) if the function receives invalid variables. The function can contain a return variable such as Error, which indicates that the function could not perform the task due to parameter errors. The function must ensure that no invalid values are returned.

It is recommended that a function only has the amount of PLC code that can be seen on the screen during the programming - 20 - 25 program code lines. If the PLC code is longer, there will be a need to create another function or try to reduce the number of PLC code lines.

It is not a technical requirement that the program code in a function is written in ST.

A function can use other functions as well as timers and counters. If a function uses timers and counters, it must be created as a function block. If the function block uses timers, the function block must be called in each program scan.

IMPORTANT: A function must never make a function call to itself!

When coding a function, there are two methods to be followed:

A) The Top-down design method

First, find a useful name for the function. Use a name that indicates what the function does. Then create a list of variables the function needs in order to solve the programming task, and a list of the variable(s) the function has to return. The list should include requirements for the variables, their data type and the variable value range (minimum and maximum values).

It is advantageous to use a variable data type which makes it easy for those who use the function. Conversion between data types has to be carried out inside the function. This is for example relevant if the function using a **TIME** data type, where it may be better to use a **WORD** data type instead, because the **TIME** data type is composed of numbers and letters.

It can give a good overview, to draw a block diagram of the variables:

Write down what the function should do. This description can be used as comments in the program code later on making it worthwhile doing.

B) The Bottom-up design method

Using this method, you start by writing individual pieces of program code. When the program code is working well, it is moved bit by bit to a function or a function block using method A.

If you do not have much experience with writing program code for functions and function blocks, the bottom-up design method is good to use.

TIP: If some of your code lines are copy-pasted and then adapted, you often need to move the code to a function!

REMEMBER: A function can be seen as a black box. When the function works well, you don't have to think too much about what's going on inside the box.

Program structure of a function

To obtain a good structure, it is recommended to follow this guide:

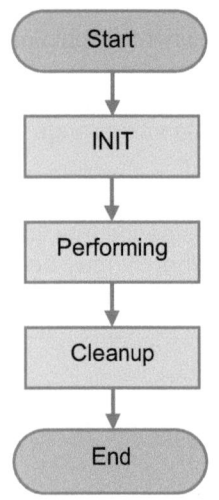

Start — The program execution starts inside the function.
The local variables are created in memory.

INIT — The first task of the function is to ensure that the variables that were included in the function call are valid.
In addition, local variables must have a start value assigned.

Performing — The function performs the specific task. It may be calculations or data that need to be moved or processed.

Cleanup — As the last step, the return variables have to be assigned a value. If it is a function block, the local variable values must be updated to be ready for the next time the function is used.
A function will lose all local variable values when the function is finished.

End — The local variable values are removed from memory and lost.
The function is finished.

Following these guidelines when using functions, you will achieve:

- A better and clearer program structure.
- Code reusability and fewer program errors (program bugs).
- The potential of speedier future expansion of the program.
- The ability to easily perform part tests on the program.
- The ability to easily disable Functions during debugging.

There are many possibilities to create useful functions. Below is a list for inspiration:

- Conversion between units of measurement from sensors.
- Calculation of expected time until next service for motors and pumps.
- Calculation of conveyor belt speed.
- Calculation of OEE (Overall Equipment Efficiency).
- Alarm monitoring of machine components.
- Code which can be reused in other programs.
- Estimation of expected production time.

The following pages contain examples of functions.

10.4.1   EXAMPLE: FC for conversion of temperature

This example shows an implementation of a function converting temperatures from Celsius (Centigrade) units to Fahrenheit units. It is implemented in a function as it is a mathematical calculation being reused many times in the program.
A **FUNCTION**, named **fcTemperatureCalculateCtoF**, is created to return a **REAL** variable, as calculations with temperature are often decimal numbers. The name of the function starts with **'fc'** to show that it is a function. The rest of the name is chosen, because it suits the task of the function. The name begins with a noun: **Temperature** and a verb: Calculate and the letters: **C to F** which indicates that the func-tion undertakes a conversion - A Celsius (Centigrade) temperature is converted to a Fahrenheit temperature.

The function has one single input parameter named **TemperatureC** which is created in the **VAR_INPUT** section as shown below. It is of the data type **REAL**, because the parameter (Celsius/Centigrade temperature) is a decimal number:

```
FUNCTION fcTemperatureCalculateCtoF : REAL
VAR_INPUT
    TemperatureC: REAL;
END_VAR
```

The PLC code inside the function is shown below:

```
//This function converts a Celsius temperature
//to a Fahrenheit temperature
//Input parameter REAL is in Celsius
//Out parameter REAL is in Fahrenheit
fcTemperatureCalculateCtoF:= (TemperatureC * 9.0/5.0) + 32;
```

The formula used for the conversion is found on the Internet.

The return parameter is the name of the function with the data type **REAL**. Comments in the beginning of the function are made to explain to other programmers what the function does. Good programming always includes comments in the beginning of the function, even if the function name is self-explanatory.

Below is shown how the function can be used, where **TempF** is a **REAL** datatype:

```
TempF :=  fcTemperatureCalculateCtoF(23.6);
//The value is copied to TempF and is 74.48 (REAL data type)
```

The function is 'called' with the value of 23.6 (Celsius/Centigrade temperature degrees C). The function returns the calculated Fahrenheit value in the variable **TempF**.

To test the function find a temperature calculation website on the Internet. Test large and small values in the function, and check if you get the same result using the website. It is always important to test functions thoroughly, because it can be difficult to find errors as programs increase in size and are made up by many functions and program modules.

10.4.2 EXAMPLE: FC to calculate average

The following chapter shows an example of a function, which calculates the average pressure of two sensors. There is no need to save values, so a **FUNCTION** is created. The function is named **ValueAverage** with two input parameters: **Value1** and **Value2**; both of the data type **REAL**. Even though it is an average of two pressure measurements, one single function is created with a general name, so that the function can be reused.

The calculated value is of the data type **REAL**, and this is defined as a return parameter for the function by writing **REAL** in the first line of the code as shown:

```
FUNCTION ValueAverage : REAL //REAL is the return parameter data type
VAR_INPUT
    Value1 :        REAL; //Input parameter 1 to the function
    Value2 :        REAL; //Input parameter 2 to the function
END_VAR
VAR
    Sum :           REAL; //Local variable for temporary calculation
END_VAR
Sum:= Value1 + Value2;  //Total sum
Sum:= Sum/2; //Average
ValueAverage:= Sum;  //Set the return parameter
```

In the last line the return value is set. This means that the **ValueAverage** is assigned a value before the function ends which ensures that the calculated value can be used outside the function. The variable **sum** is a local variable and can therefore not be used outside of the function. This creates a good program structure and is a 'black box' to calculate an average of two values.

There are several different ways to use the **ValueAverage** function. The variables below: **Avg1**, **Avg2**, **Avg3**, **Sensor1Pressure**, **Sensor2Pressure** are all of the data type **REAL**, because it is the data type which the **ValueAverage** function uses as input parameters and return parameter.

Examples of how to use the **ValueAverage** function:

```
//Example #1: Use the function variable names
Avg1:= ValueAverage(Value1:= 85.1, Value2:= 17.6);
//Example #2: Use value only
Avg2:= ValueAverage(85.1, 17.6);

//Assign values to main variables
Sensor1Pressure:= 85.1;
Sensor2Pressure:= 17.6;

//Example #3: Use main variables
Avg3:= ValueAverage(Sensor1Pressure, Sensor2Pressure);
//Example #4: Combination of #2 and #3
Avg4:= ValueAverage(Value1:= Sensor1Pressure, Value2:= Sensor2Pressure);
```

When a **FUNCTION** is used, all input parameters must have a value. When using a **FUNCTION_BLOCK** not all input parameters require a value. However, it is a good idea to assign all input parameters a value, because it indicates that the programmer knows how to use parameters and remembers them.

The order of which parameters are 'called' in the function is important and remains the same as when the function was created. Therefore, **Value1** has to be written first followed by **Value2**.

ValueAverage is a function and therefore it is not necessary to write the parameters **Value1:=** and **Value2:=** as shown at **#1** and **#4**.

The PLC code inside a function *MUST* take into consideration whether input parameters are missing, invalid or parameters are outside the permitted range. The PLC code inside the function must be stable and able to be executed, even if input parameters are missing, are wrong or invalid.

Furthermore, the programmer, who uses the function must also ensure that the function is 'called' with valid input parameters. The programmer must write a description of the function explaining the mode of operation and input parameters. Most importantly, the function is not ready to be used before it has been tested!

10.4.3 EXAMPLE: FC for level measurement in tank

This example shows how a function can be used to check if a signal is inside a measurement range. As shown on the diagram below, the liquid level in a tank is divided into different measuring ranges. A light turns on when the level in the tank reaches a specified range.

A function is used to check whether values are inside measurement ranges, which makes it easy to reuse the code. If a function was not used here, the same PLC code lines would have to be written three times, which is bad program structure.

The illustration below shows the connection between components, PLC and code:

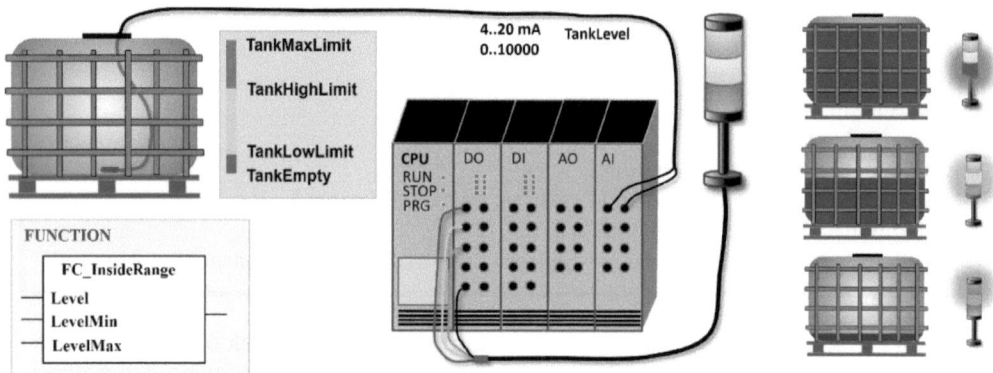

The PLC code is shown on the next page, where **CONSTANT**s are used to define the different limits of the measurement ranges. When using **CONSTANT** the limit values are easy to change as they are only required to be made in one place.

The signal from the submerged level sensor inside the tank has a 4-20 mA signal, and is connected to the analog input card (AI). In this example, the analog signal is scaled to a range from 0 to 10000 and therefore a **WORD** data type is used.

The function uses the relational operator **> =** (greater than) when comparing with **LevelMin**, and therefore 1 must be added to **TankLowLimit** and **TankHighLimit** before calling the function, otherwise two lights will be turned on when levels are exactly 1000 or exactly 7000.

```
FUNCTION FC_InsideRange : BOOL
VAR_INPUT
  Level:      WORD;
  LevelMin: WORD;
  LevelMax: WORD;
END_VAR
```

```
//The FUNCTION returns TRUE if Level is
// inside MIN and MAX range. If not inside range FALSE is returned
IF Level >= LevelMin AND Level <= LevelMax THEN
  FC_InsideRange:= TRUE;
ELSE
  FC_InsideRange:= FALSE;
END_IF;
```

```
PROGRAM MAIN
VAR
  L1_Red:     AT %IX0.0 BOOL := FALSE; //OUTPUT Connection to Lamp
  L2_Yellow: AT %IX0.1 BOOL := FALSE; //OUTPUT Connection to Lamp
  L3_Green:  AT %IX0.2 BOOL := FALSE; //OUTPUT Connection to Lamp
END_VAR
VAR CONSTANT
  TankMaxLimit: WORD := 10000;
  TankHighLimit: WORD := 7000;
  TankLowLimit: WORD := 1000;
  TankEmpty:     WORD := 0;
END_VAR
VAR
  TankLevel: AT %IW3 WORD := TankEmpty; // Analog sensor. Range 0 - 10000
END_VAR
```

```
//Main PLC program code
L1_Red:=    FC_InsideRange (TankLevel, TankEmpty, TankLowLimit);
L2_Yellow:= FC_InsideRange (TankLevel, TankLowLimit + 1, TankHighLimit);
L3_Green:=  FC_InsideRange (TankLevel, TankHighLimit + 1, TankMaxLimit);

//This code can be used as test code, when there are no sensor connected:
//TankLevel:=TankLevel + 1;
```

10.4.4   EXAMPEL: FC to linear scaling of sensor signal

This chapter describes a function that can be used for linear scaling of an analog sensor signal. Normally this type of function can be found in the PLC progam library, or the function exists directly on input or output card.

This example will help you learn and understand the mathematics behind linear scaling and the construction of a function.

Scaling of values is needed when a value needs to follow a different scale. The example to the right involves a sensor value from the measuring range of 4-20 mA, which must be scaled to a measuring range of 0 to 100%.

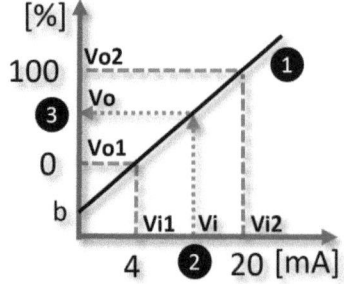

The linear scaling method consists of a straight line (1) in a coordinate system as shown on the diagram.

A value (2) from the signal measuring range [mA] on the x-axis corresponds to a specific value (3) in percentage on the y-axis.

A linear scale uses the formula of a straight line: $y = ax + b$

Where the slope is calculated using the formula: $a = (Vo2 - Vo1) / (Vi2 - Vi1)$

The point of intersection on the y-axis b is found: $b = y - ax$ => $b = Vo2 - a * Vi2$

The function must take **Vi1**, **Vi2**, **Vo1** and **Vo2** as input parameters because the values are used in the calculation, and this makes the function general. The value to be scaled is **x** (shown as **Vi**). The value to be calculated is **y** (shown as **Vo**).

When designing and writing the code for the function, the variable names: **x**, **y**, **Vi1**, **Vi1**, **Vo1** and **Vo2** will be replaced with meaningful variable names. For internal calculations, the variable names **a** and **b** are used, because they are used in the formulas, and it will therefore be easier for other programmers to understand the calculations.

It is very important that the function checks that the input parameters are inside the valid range, because invalid values can result in an unstable program. Therefore, the first code lines inside the function checks whether the input values are valid. If input values are invalid, the variable **Error** ensures that the output value is set to zero (0).

Lastly, the function is tested with different input parameters to make sure the function works as expected. This is called program module testing.

```
FUNCTION Scale : REAL
VAR_INPUT
  ValueIn:      REAL; // (Vi) Value to be scaled
  ScaleInMin:   REAL; // (Vi1) Scale in Min value. Must be lower than ScaleInMax
  ScaleInMax:   REAL; // (Vi2) Scale in Max value. Must be higher than ScaleInMin
  ScaleOutMin: REAL; // (Vo1) Scale out Min. Must be lower than ScaleOutMax
  ScaleOutMax: REAL; // (Vo2) Scale out Max. Must be higher than ScaleOutMin
END_VAR
VAR
  a:     REAL;              // Slope of the curve
  b:     REAL;              // Intersection with y-axis
  Error: BOOL := FALSE;    // Input value error. TRUE if error
END_VAR
```

```
//First check that input values are valid
IF ScaleOutMin >= ScaleOutMax THEN
  Error := TRUE;
END_IF;

//Check value to avoid division by zero
IF ScaleInMin >= ScaleInMax THEN
  Error := TRUE;
END_IF;

//Check valueIn is inside range
IF ValueIn < ScaleInMin OR ValueIn > ScaleInMax THEN
  Error := TRUE;
END_IF;

//Perform calculation if no error
IF Error = FALSE THEN
  a:= (ScaleOutMax - ScaleOutMin) / (ScaleInMax - ScaleInMin);
  b:= ScaleOutMax - (a * ScaleInMax);
  //Set output value: y=ax + b
  Scale:= a * ValueIn + b; //Return value
ELSE
  Scale:= 0; //Return zero (0) in case of input value error
END_IF;
```

Below the function is tested with different input values. Both invalid and valid input values are used to ensure that the function works as expected:

```
Value0 := Scale(12, 4, 20, 0, 100);     //Expected result: Value0 = 50
Value1 := Scale(-2, 4, 20, 0, 100);     //Expected result: Value1 = 0
Value2 := Scale(20, 4, 20, 0, 100);     //Expected result: Value2 = 100
Value3 := Scale(6, 20, 4, 100, 1100); //Expected result: Value3 = 0
Value4 := Scale(5, 4, 20, 100, 1100); //Expected result: Value4 = 162.5
```

11 Working with text and chars, STRING

STRING is the data type to be used, when working with text. Shown below are some areas where a PLC uses text and characters (char):

Showing dynamic texts and digits on HMI (Human Machine Interface):

- Online changes between languages on a user operation panel (e.g. switch between Danish and English language user interface with no changes to the PLC code) (multi user language change)
- Messages and instructions to the user: production information, typing in passwords, reading of letters, time/date, alarm texts

Handling files and database data:

- Reading data from files on a hard disk (e.g. settings of equipment and instrumentations, configuration files, set points)
- Data logging of data or event measurements (e.g. changing settings or mechanical condition changes)
- Texts read from hard disk or flash card
- Messages to/from production systems (ERP, SAP, MES, WCS)
- File names, folder names, e-mail

Data communication between PLC/PC/Instruments:

- Instruments send data in ASCII (e.g. BAR/QR codes, RFID, TAGS)
- Information to label printer (e.g. labels to boxes, production dates)
- SMS (e.g. alarms/commands to/from mobile phones)
- Numbers with many digits mixed with letters
- Data measurements, alarms, information from automation equipment

The following lists the data types dealing with text:

Data type	Description
CHAR	Contains one character only (ASCII) (8 bit)
WCHAR	Contains a wide character (16 bit) (UNICODE, ISO 10646)
STRING	ARRAY of CHARS [0..254], for sentences (254 is max.)
WSTRING	ARRAY of WCHAR [0..254], for sentences (254 is max.) Used for PLC controls handling multiple languages on HMI (Human Machine Interface) (UNICODE, ISO 10646)

NOTICE:

> Only use **STRING** when necessary, because it requires CPU power and uses a lot of memory.
>
> Only create **STRING** with the array length needed.

Not all PLC types provide the data types **CHAR** and **WCHAR**. If a variable only has one single sign (character), create a **STRING[1]** or a **BYTE**.

> **IMPORTANT:** The length of a **STRING** is defined by counting characters until element 0 (zero) is found in the **ARRAY** (some programming languages puts the length of the **STRING** at element zero, which is important to know if a PLC communicates with other equipment)

A **STRING** shows characters by using an ASCII table. These are saved as integers in an **ARRAY**, because a CPU is only able to save data in integers. Below is shown an **ARRAY** with integers and the corresponding characters from the ASCII table:

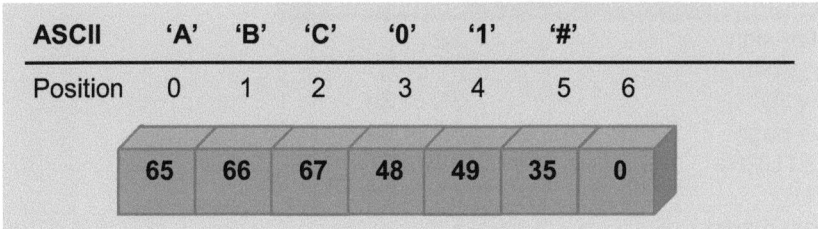

A PLC typically provides a maximum length of 255 characters in a **STRING**. If a text is longer than 255 characters, the text can be split up into several **STRINGs**.

A **STRING** can be created with or without a fixed length as shown below:

```
PROGRAM DemoString
VAR
   szDemo:    STRING     := 'No fixed length';
   szDemoFix: STRING[35] := 'Fixed length string';

   szEmpty:   STRING     := '';        //String without text
   szDemoW:   WSTRING    := " This is a UNICODE string "; //Text with 2 x "
END_VAR
```

If NO length is indicated – as is the case for **szDemo** – the PLC uses 254 bytes in the memory + 1 (zero sign for ending the **STRING** is included).

If a fixed length is set – as is the case for **szDemoFix** – the PLC uses the fixed length – in this example 35 bytes of the memory + 1 (zero for ending is included).

The above indicates that the best choice is to set a maximum length on all **STRINGs**. However, as texts can be dynamic during the execution of a program, challenges can arise from using a fixed **STRING** length. This can e.g. be the case when making language changes online, where texts can be 50 % longer, when changing from an English text to a French text.

It is not possible to write text with double citation sign: *A "big" test*. An escape character must be placed ($ sign) before the text: A *$"big"* test.

Possible escape sequences:

Description	Sequences
Dollar sign	$$
Line shift	$L or $l
New line	$N or $n
New page	$P or $p
<RETURN>	$R or $r
<TAB>	$T or $t
Citation sign	$'
Double citation sign	$" or $22

11.1 EXAMPLE: FC with STRING

Below is shown an example a **FUNCTION** with **STRING**:

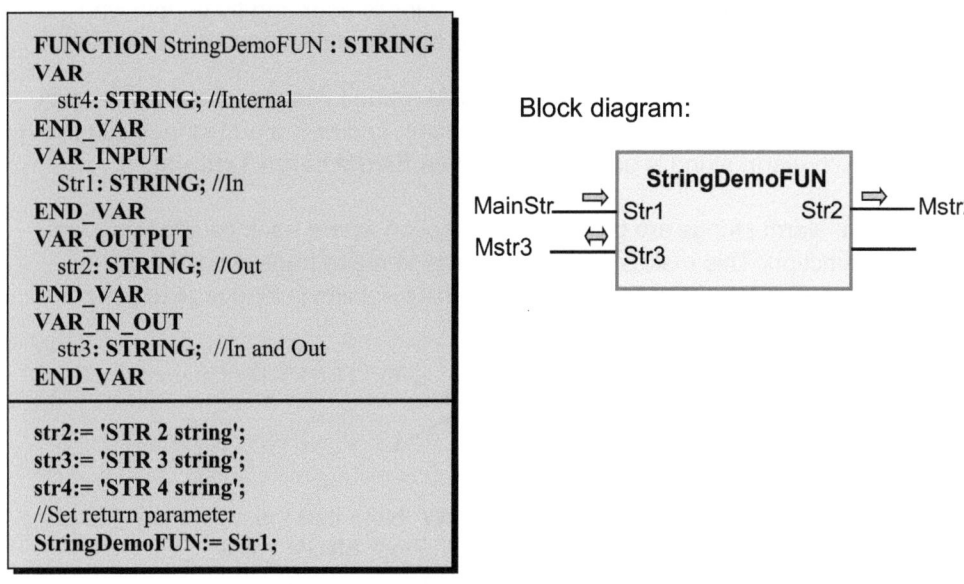

```
FUNCTION StringDemoFUN : STRING
VAR
    str4: STRING; //Internal
END_VAR
VAR_INPUT
    Str1: STRING; //In
END_VAR
VAR_OUTPUT
    str2: STRING;  //Out
END_VAR
VAR_IN_OUT
    str3: STRING;  //In and Out
END_VAR

str2:= 'STR 2 string';
str3:= 'STR 3 string';
str4:= 'STR 4 string';
//Set return parameter
StringDemoFUN:= Str1;
```

Block diagram:

Program 'call' to the function **StringDemoFUN:**

```
PROGRAM Main
VAR
    MainStr, Mstr1, Mstr3: STRING;
    Mstr2: STRING [5]; //Limited to 5 chars
END_VAR

MainStr:= 'Hello World';
Mstr1:= StringDemoFUN (str1:=MainStr, Str2=>Mstr2, str3:=Mstr3);

//Contents of the variables are:
//Mstr1 = 'Hello World'.
//Mstr2 = 'STR 2'   //Because STRING length is 5: Mstr2[5]
//Mstr3 = 'STR 3 string'.
```

NOTICE:
The variables **Mstr1**, **Mstr2** and **Mstr3** are all created with a **STRING** data type.
As the variable **Mstr2** is created with a fixed length of 5, it will only contain 5 characters, even if the string **str2**, used inside the function contains 12 characters.

11.2 EXAMPLE: Program structure for language change

This chapter consists of a proposal for a program structure, which allows alarm texts to be displayed in different languages. If the machine is operated by people who speak different languages, it is an advantage to enable online language switching.

The various alarm texts are declared in an **ENUM** named **Alarms**. Each alarm texts has a unique number through the ENUM indexation, and new alarm strings can easily be added. Each alarm string is also declared in an **ENUM** called **TxtLang**.

The individual alarm strings are grouped in a function where each country language has its own function. This makes it clear and easy to make translations into other languages. The **GetAlarmTxt** function is used to select which language to use:

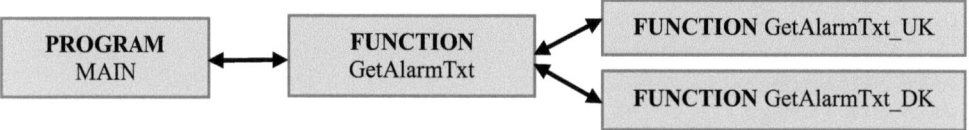

In the program, alarms are declared in an **ARRAY**. Note that the maximum size of **ARRAY** is **ALARMS_MAX**, and is definded as the last **ENUM** value.

The current language is selected by the variable **TxtLanguage**.

```
PROGRAM MAIN
VAR
  MotorAlarmArr:  ARRAY [0..Alarms.ALARMS_MAX] OF BOOL;
  AlarmTxt:       WSTRING; //Alarm text to be shown
  TxtLanguage:    INT := TxtLang.UK; //Language to be used
END_VAR

//Set overload alarm
MotorAlarmArr[Alarms.OVERLOAD]:= TRUE;

//Reset temperature alarm
MotorAlarmArr[Alarms.TEMPERATURE]:= FALSE;

//Get the alarm text for OVERLOAD
IF MotorAlarmArr[Alarms.OVERLOAD] THEN
  AlarmTxt:= GetAlarmTxt(TxtLanguage, Alarms.OVERLOAD);
END_IF;
```

The below contains example code for the **ENUM** and functions:

```
TYPE Alarms :
  ( NONE_ALARM, TEMPERATURE, NO_OF_STARTS, OVERLOAD, ALARMS_MAX );
END_TYPE
```

```
TYPE TxtLang :
  ( None, UK, US, DK , DE, ES);  // Country codes
END_TYPE
```

```
FUNCTION GetAlarmTxt : WSTRING
VAR_INPUT
  Language: WORD;
  AlarmNo: WORD;
END_VAR
VAR_OUTPUT
  AlarmTxt: WSTRING;
END_VAR
VAR
  Str : WSTRING;
END_VAR
```

```
//Program code for GetAlarmText
CASE Language OF
  TxtLang.UK : Str:= GetAlarmTxt_UK(AlarmNo);
  TxtLang.DK : Str:= GetAlarmTxt_Dk(AlarmNo);
ELSE
  Str:= "Unknown langauge selected";
END_CASE;

//Set return value
GetAlarmTxt:= Str;
```

```
FUNCTION GetAlarmTxt_UK :
                       WSTRING
VAR_INPUT
  // Get Alarm Text for the AlarmNo
  AlarmNo: WORD;
END_VAR
VAR
  // The alarm text for the AlarmNo
  Str: WSTRING;
END_VAR
```

```
//Program code for GetAlarmTxt_UK
CASE AlarmNO OF
  Alarms.NONE_ALARM    : Str:= "No Alarms";
  Alarms.TEMPERATURE : Str:= "A1 Motor Temperature";
  Alarms.NO_OF_STARTS : Str:= "A2 Too many starts";
  Alarms.OVERLOAD        : Str:= "A3 Motor overload";
ELSE
  Str:=  "Alarm Text not found";
END_CASE;

 //Set retun value
GetAlarmTxt_UK:= Str;
```

```
FUNCTION GetAlarmTxt_DK :
                       WSTRING
VAR_INPUT
  // Get Alarm Text for the Alarm No
  AlarmNo: WORD;
 END_VAR
VAR
  // The alarmtext for the AlarmNo
  Str: WSTRING; for the AlarmNo
END_VAR
```

```
//Program code for GetAlarmTxt_UK
CASE AlarmNO OF
  Alarms.NONE_ALARM   : Str:= "Ingen alarmer";
  Alarms.TEMPERATURE : Str:= "A1 Motor temperatur";
  Alarms.NO_OF_STARTS : Str:= "A2 For mange motor start";
  Alarms.OVERLOAD        : Str:= "A3 Motor overbelastet";
ELSE
  Str:= "Alarm tekst ikke fundet";
END_CASE;

//Set return value
GetAlarmTxt_DK:= Str;
```

11.3 Standard functions, STRING

The built-in standard **STRING** functions are shown below. Some PLC types provide more functions which can be found in the manufacturer's programming manual.

If a different **STRING** function is needed, the programmers have to write the code and implement it themselves, or try to find a function on the internet.

The max length for **STRING** in the standard functions is 255 characters.

CONCAT

Connects two **STRING**
STR2 is inserted after **STR1**

Str3:= **CONCAT** (**STR1** := 'AB', **STR2**:='CD');
//Str3 = 'ABCD'   alternatively use: Str3:= **CONCAT** ('AB', 'CD');

INSERT

Inserts a **STRING** in another **STRING** at a certain position. **STR2** is inserted in **STR1** at **POS** position

Str3:= **INSERT** (**STR1**:='ABCD', **STR2**:='EFGH', **POS**:=2);
//Str3 = 'ABEFGHCD'   Str3:= **INSERT** ('ABCD', 'EFGH', 2);

DELETE

Delete some part(s) of a **STRING**. **IN1** is the **STRING**
From position **POS** the amount, which **LEN** indicates, is deleted

Str3:= **DELETE** (**IN1**:='ABCDEFG', **LEN**:=2, **POS**:=3);
//Str3 = 'ABEFG'   Str3:= **DELETE** ('ABCDEFG', 2, 3);

REPLACE

Replaces some parts(s) of a **STRING**. **L** characters in **STR1** is deleted.
STR2 is inserted from position **P**

Str4:= **REPLACE** (**STR1**:='ABCDEFG', **STR2**:='X', **L**:=2, **P**:=3);
//Str4 = 'ABXEFG'   Str4:= **REPLACE** ('ABCDEFG', 'X', 2, 3);

FIND

Find a **STRING** in another **STRING**. A match for **STR2** is searched for in **STR1**.
An **INT** is returned with the position where **STR2** was found in STR1.
If nothing is found, 0 (zero) is returned. The **FIND** function is case sensitive, i.e. it
distinguishes between upper case and lower case letters.

```
Int1:= FIND (STR1:='ABCBCDEFG', STR2:='BC');
//Int1 = 2  'BC' is found first at position 2   Int1:= FIND ('ABCBCDEFG', 'BC');
```

LEN

LEN finds the length of a **STRING**. Counting numbers of characters in **STR**
An **INT** with the length is returned.

```
Int2:= LEN (STR:= 'Demo');
//Int2 = 4              alternatively use:        Int2:= LEN ('Demo');
```

LEFT

LEFT keeps some part(s) of a **STRING** starting from the left.
The first parameter **STR** is **STRING** and the second parameter **SIZE** is the
number of characters which is counted.

```
Str6:= LEFT(STR:='1234567', SIZE:=2);
//Str6 = '12'          alternatively use:        Str6:= LEFT( '1234567', 2);
```

RIGHT

RIGHT keeps some part(s) of a **STRING** starting from the right.
The first parameter **STR** is **STRING** and the second parameter **SIZE** is the
number of characters which is counted.

```
Str6:= RIGHT (STR:='1234567', SIZE:=2);
//Str7 = '67'          alternatively use:        Str7:= RIGHT( '1234567', 2);
```

MID

MID keeps some part(s) of a **STRING**.
The first parameter **STR** is **STRING**, **LEN** is the length of what will be retained,
and **POS** is the starting position of what will be retained.

```
Str7:= MID (STR:='1234567', LEN:=2, POS:=3);
//Str8 = '34'                alternatively use: Str8:= MID( '1234567', 2, 3);
```

As **STRING** is an **ARRAY**, not all PLC types support the use of relational operators (see chapter 7.2, page xx, side 41) directly on a **STRING**. The built-in **FIND** and **LEN** functions must be used when comparing texts:

```
Str1 := 'abc';
Str2 := 'abc';

IF Str1 = Str2 THEN
  Str3:= 'Ens';
END_IF;
```
X

```
Str1 := 'abc';
Str2 := 'abc';

IF FIND (Str1, Str2) > 0 THEN
  IF LEN (Str1) = LEN (Str2) THEN
    Str3:= 'Ens';
  END_IF;
END_IF;
```
OK

For converting numbers, the built-in data type conversion functions can also be used on **STRINGs** (see chapter 8.6, page 51) as seen below:

```
myInt:= STRING_TO_INT('123');
myReal := STRING_TO_REAL ('12.45');
myStr1 := REAL_TO_STRING (23.67);
```

Before conversion functions are 'called', the string (which is the input parameter) must be checked to ensure the function does not receive characters in a string which are not convertible. It might be unclear what will happen if the PLC program converts e.g. 'ABC' to a **REAL** data type. You can find functions and function code on the internet that can be used to check whether the contents of a string is a number. The **IsNumber** function is an example of this, and can be found via a google search.

IMPORTANT: In some PLC types, **STRING** standard functions are not 'thread safe'. This means that the best choice is to only make use of them in PLC code being executed in the same PLC-task.

Some PLC types support functions to handling wide strings: Like **WCONCAT** or **WLEN**.

As **STRING** is an **ARRAY**, it is possible to insert a character directly into it. Below three different examples are shown, as different PLC types handle this differently:

```
str1:= 'My String';
str1[2]:= 'A';           //Solution 1, insert 'A' into location 2 in str1
str1[2]:= 65;            //Solution 2, where 65 is 'A' in the ASCII tabel
str1[2]:= F_toASC('A');  //Solution 3, use a built-in function named F_toASC

//The resulting string is 'MyAString' where 'A' is overwriting <SPACE> in str1
```

11.4 EXAMPLE: FC Find numbers in a STRING

This example describes a function that can be used to find a number in a particular position in a **STRING** containing numbers separated by a semicolon (;):

'10;30;45;200;4;5;3;4;23;30;90;8;65'

The function is used in the example found on page 166, where each number in the STRING is used to execute sub-programs for a car wash. The user can enter the sub-programs that a complete car washing program should include.

The example here also shows how to use the **STRING** standard functions.

A block diagram of the function and parameters used is shown below:

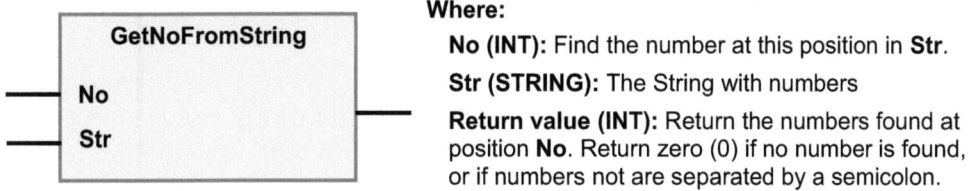

Where:

No (INT): Find the number at this position in **Str**.

Str (STRING): The String with numbers

Return value (INT): Return the numbers found at position **No**. Return zero (0) if no number is found, or if numbers not are separated by a semicolon.

A **FUNCTION** is used to ensure that the program code can easily be reused.

Program example for using and calling the function:

```
//no1, no2 and no3 are INT datatype
no1:= GetNoFromString (2, '1;10;12;11;10;10;10');   //Return: no1 = 10 (at position 2)
no2:= GetNoFromString (4, '7;10;13;14;10;15;10');   //Return: no2 = 14 (at position 4)
no3:= GetNoFromString (6, '3;11;120;43;20');        //Return: no3 = 0, because not found
```

The program code can be found on the next page. The code is split up into two sections: The first section verifies that the **STRING** only contains numbers and semicolons, and ensures not continue in case of a non-valid **STRING**. Verification is performed by comparing each character, with characters in the ASCII table. The character semicolon (';') has number 59 in the ASCII table, and the numbers between 48 and 57 are all found in the ASCII table.

The **PosEnd** variable points to the semicolon after the number to be returned.

The last section of the program finds the number in **Str**. The number can be up to 3 digits long. Finally, the returned number is converted from a **STRING** to an **INT**

```
FUNCTION GetNoFromString : INT
VAR_INPUT
  No:        INT;        // Get number. Return zero (0) if nothing is found
  Str:       STRING; // The current string can be '23;4;34;100;2;60'
END_VAR
VAR
  StringOk:  BOOL := FALSE; // valid if STRING it contains numbers and ';'
  i:         INT;        // Counter for the FOR loop
  PosEnd:    INT;        // Pointer to the ';' after the number found
  SepNO:     INT := 0; // Count no of ';' in the STRING
END_VAR
```

```
//Input string can be '1;3;7;10;101;12;1;' or '234;3;78;8;43;100'

Str:= CONCAT(str,';'); //Place ';' after last number
Str:= INSERT(str,'00', 0); //To allow more than 2 digits
GetNoFromString:= 0; //Sets 0 as return value if string not valid

FOR i:= 0 TO LEN(Str) DO //Consider each sign in the STRING
  IF Str[i] = 59 OR (Str[i] >= 48 AND Str[i] <= 57) THEN //ASCII ';' or 0..9
    StringOK:= TRUE; //Text ok
    IF Str[i] = 59 THEN //ASCII check is ';' found
      SepNO:=SepNO + 1; //Number of ';' in the STRING
      IF SepNO = No THEN //Position found
        PosEnd:= i + 1;  //Where no is ended
      END_IF; //SepN0
    END_IF; //str[i]
  END_IF;
END_FOR;

IF StringOK = TRUE THEN //If STRING ok, get value
  Str:= MID(Str, 3, PosEnd - 3); //Number has max 3 digits
  i:= FIND(Str,';'); //Any ';' found?
  IF i = 2 THEN //If like '2;3' => '003'
    Str[1]:= 48; Str[0]:= 48; //Place '00'
  END_IF;
  IF i = 1 THEN //If like ';23' => '023'
    Str[0]:= 48; //Place '0'
  END_IF;
  GetNoFromString:= STRING_TO_INT(str); //Return value
END_IF;
```

The **GetNoFromString** function contains 29 lines of code including comment lines. This is a lot of lines, and more than the recommended 20-25 lines of code for a readable and clear program structure. Sometimes it is not possible to reduce the number of code lines to keep a good program structure, but this part of the program code:

```
i:= FIND(str,';'); //Any ';' found?
IF i = 2 THEN //If like '2;3' => '003'
  Str[1]:= 48; Str[0]:= 48;  //Place '00'
END_IF;
IF i = 1 THEN //If like ';23' => '023'
  Str[0]:= 48; //Place '0'
END_IF;
```

Can be reduced to this, where **CASE** is used instead of **IF** statements:

```
CASE FIND(str,';') OF //Any ';' found?
  2 : Str[1]:= 48; Str[0]:= 48; //If like '2;3' => '003'
  1 : Str[0]:= 48; //If like ';23' => '023'
END_CASE;
```

The code now consists of 4 lines instead of 7 lines. In addition, the variable **i** is no longer used in the program code. The variable **i** is used earlier in the function, as it is also used as a counter in the **FOR** loop.
It is possible to reuse variables, to reduce the number of variables which need to be created. However, reusing variables can also make the code more unreadable.

If using **STRING** functions and **CASE** statements, the program code looks like this:

```
CASE FIND(str,';') OF //Any ';' found?
  2 : Str:= REPLACE(Str, '00', 2, 1); //If like '2;3' => '003'
  1 : Str:= REPLACE(Str, '0', 1, 1); //If like ';23' => '023'
END_CASE;
```

If only **STRING** functions are used, the code will look like this:

```
Str:= RIGHT (Str, LEN(Str) - FIND (Str,';')); //Remove any ';'
```

The point is that the four sections of program code all work in the same way, but they use different methods and functions. It is important to write code, so other programmers can understand it, and easily change the code at a later stage.

11.5 FB: Optimize insertion of values into STRUCT

Example page 166, shows how values can be inserted into an **ARRAY** and a **STRUCT** to configure different washing programs. If you want to insert many values, the code can quickly become very long which is not good program structure.
The program code to insert values into **ARRAY** and **STRUCT** looks like this:

```
ArCarWash [1].ProgramName := 'Budget Wash';
ArCarWash [1].ProgramNumbers := '1;11;20;';
ArCarWash [1].Cost:= 10;
```

Optimization of the above program code can be carried out by using **FUNCTION**.

There are four different values: **[1]**, **ProgramName**, **ProgramNumbers** and **Cost** and therefore a function with four input parameters must be used. It is important that the function checks that the input parameters are inside the valid range, because invalid parameters can cause the program execution to stop.

To create a good program structure, the **ArCarWash** variable is also an input parameter to the function.This means that function variables are not written directly to a variable in the program module. A function can write directly to a program module by changing **VAR** to **VAR_IN** for the **ArCarWash** variable in the program module where **ArCarWash** is declared.

ArCarWash is an **ARRAY** and it is therefore only the address which is transferred to the function using the variable scope **VAR_IN_OUT**.

The **NoError** variable is used to ensure that values are only written to **ArCarWash** if they are valid. The variable is set to **FALSE** each time the function is called and does not change if an error is found with one of the parameters. The return value can be used to notify the programmer/operator that input parameters were invalid.

The internal variables **IndexMin**, **IndexMax** and **CharTemp** are used to provide a good program structure and ensure that code lines are not too long.

The program call for the function is:

```
FCWashPrgConfig(ArCarWash, 1, 'Budget Wash', '1;11;20;', 10);
```

By using a function, the program is reduced from three lines of code to a single line. In addition, the parameters are now also checked to see if they are inside the valid value range.
See the next page to view the program code for the function:

```
FUNCTION FCWashPrgConfig : BOOL
VAR_IN_OUT
 WashArray: ARRAY [*] OF CarWashType; //Pointer to the ARRAY which can have any size
END_VAR
VAR_INPUT
 Index:      WORD;    //Index to the WashArray
 Name:       STRING;  //Name to be inserted
 Numbers:    STRING;  //Numbers in a STRING to be inserted
 Cost:       REAL;    //Cost to be inserted
END_VAR
VAR
 i:                    WORD;  //For the FOR loop
 IndexMax, IndexMin:   WORD;  //Max and Min Index of the WashArray
 NoError:              BOOL := FALSE; //Set default to FALSE
 CharTemp:             STRING; //One char found in the Number STRING
END_VAR
```

```
//Program code for FCWashPrgConfig

//Get and save Index as it is used more than once in the code
IndexMin:= DINT_TO_WORD(LOWER_BOUND (WashArray,1));
IndexMax:= DINT_TO_WORD(UPPER_BOUND (WashArray,1));

//Check that Index is inside the ARRAY min and max range
IF Index >= IndexMin AND Index <= IndexMax THEN
  NoError:= TRUE; //Ok
END_IF;

//Check that "Numbers" contains valid chars
IF NoError THEN //Only perform if no error
  FOR i:= IndexMin TO IndexMax DO //For each char in the Numbers ARRAY
    CharTemp:= MID(Numbers, 1, INT_TO_WORD(i)); //Get next char in the Numbers STRING
    IF ((FIND (';0123456789',CharTemp) = 0) AND NoError) THEN //Valid char?
      NoError:= FALSE; //An error found in the Numbers STRING
    END_IF;
  END_FOR;
END_IF;

//Insert all values if there is no error
IF NoError THEN
  WashArray[Index].ProgramName:= Name;
  WashArray[Index].ProgramNumbers:= Numbers;
  WashArray[Index].Cost:= Cost;
END_IF;

//Set return to signal TRUE or FALSE. If TRUE no values are inserted
FCWashPrgConfig:= NoError;
```

12 Built-in standard functions

This chapter describes a number of built-in standard functions. When to use them depends on the task. Bear in mind that the functions can be named differently in different PLC types. This means that if the standard built-in functions are used, it can make it more difficult to copy the PLC code to other PLC types, because the code might need to be adjusted.

12.1 First program execution: First ScanBit

Some part(s) of the PLC code might need to be executed once only right after powering up the PLC (PLC turn on). It could be digital outputs that must be initialized to a certain value to ensure that e.g. signal lamps turn on with the correct color light, or a valve is set to OFF. Maybe internal variables, counters and arrays must be reset to zero at startup.

Some PLC types provide a first-scan-bit or *FirstCycleBit* for this purpose. However, if the PLC does not provide such a feature, the below PLC code can be used:

```
PROGRAM MAIN
VAR
   FirstScanBit : BOOL := FALSE; //#2
END_VAR

//Set first scan bit
IF FirstScanBit = FALSE THEN //#3
   // Initialization code here, or call to a program module
   // code here will be executed only once
   FirstScanBit := TRUE; // #1
END_IF;
```

Mode of operation:
A **BOOL** variable **FirstScanBit** is created which is initialized in the variable section to **FALSE** (see **#2**). This causes the first-scan-bit to *always* be initialized as **FALSE**, when starting up the PLC. When the PLC code is executed the first time, the PLC code within the **IF** statement will be executed as **FirstScanBit** is **FALSE** (see **#3**). When **FirstScanBit** is set to **TRUE** the PLC code **#1** is never executed again.

12.2 Edge detection (One shot): R_TRIG, F_TRIG

There is often a need for PLC code to only be executed once related to a certain action. It can be a sensor switch which is activated triggering the execution of specified PLC code (e.g. a sensor counting objects on a conveyer belt). When the sensor switch is activated, the code will be executed several times due to the mode of operation of which a PLC executes a program. Take this into account and write code to prevent multiple code executions, if you need to.

There are two standard function blocks to make sure that code is only executed once:

R_TRIG (One Shot Rising, positive edge detecting, OSR)

The function **R_TRIG** provide an input parameter **CLK** and an output parameter **Q**, both of the data type **BOOL**.
R_TRIG is used on a rising signal, where **CLK** goes from **FALSE** to **TRUE** and when it happens, **Q** is **TRUE** during one program-scan.

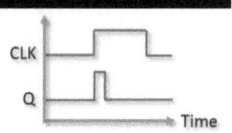

F_TRIG (One Shot falling, negative edge detecting, OSF)

The function **F_TRIG** provide an input parameter **CLK** and an output parameter **Q**, both of the data type **BOOL**.

F_TRIG is used on a declining signal, where **CLK** goes from **FALSE** to **TRUE** and when it happens, **Q** is **TRUE** during one program-scan.

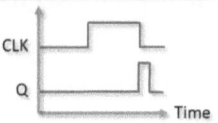

Below a program example is shown:

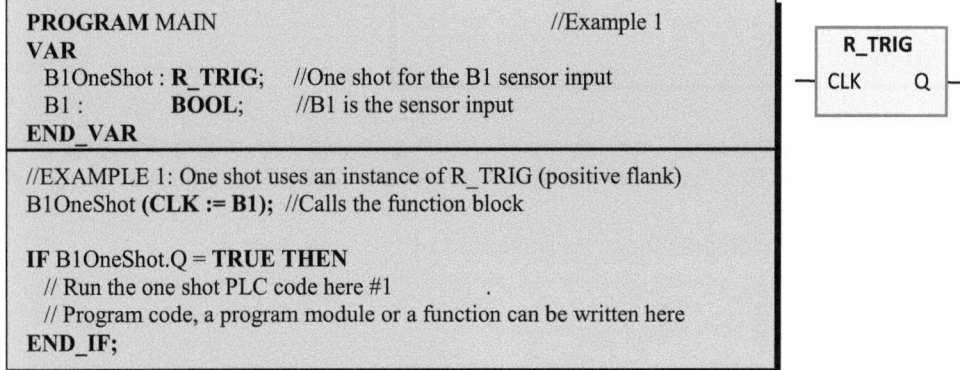

```
PROGRAM MAIN                                    //Example 1
VAR
  B1OneShot : R_TRIG;    //One shot for the B1 sensor input
  B1 :        BOOL;      //B1 is the sensor input
END_VAR

//EXAMPLE 1: One shot uses an instance of R_TRIG (positive flank)
B1OneShot (CLK := B1);  //Calls the function block

IF B1OneShot.Q = TRUE THEN
  // Run the one shot PLC code here #1           .
  // Program code, a program module or a function can be written here
END_IF;
```

The mode of operation is as follows:

B1 becomes **TRUE** when the sensor switch is activated, and **B1** is the input parameter to the **B1OneShot** function block. It sets the **BOOL** variable **B1OneShot.Q** to **TRUE** in the program-scan, when **B1** become **TRUE**.
In the following program-scan, **B1OneShot.Q** is automatically set to **FALSE** by the built-in R_TRIG function. The PLC code in #1 section is therefore only executed once.

Example 2 is a do-it-yourself solution without using **R_TRIG**. Here the physical switch is **B1** and when it is 1 (activated by e.g. a switch or a sensor used for counting objects on a conveyer belt) at the same time as **B1Old** is 0, the PLC code will be executed, marked by **#1**. When the code in **#1** is executed, **B1Old** is set to 1. In the next program-scan the code is not executed. When **B1** is 0 again, **B1Old** is set to 0.

Below the program example is shown:

```
PROGRAM MAIN                          //Example 2
VAR
   B1:     BOOL; //Sensor or switch
   B1Old: BOOL; //Internal use
END_VAR

//EXAMPLE 2: Using own PLC code
//Detect on rising edge
IF  B1 = 1 AND B1Old = 0 THEN
   B1Old := 1;
   //Insert PLC code here to run only once #1
END_IF;

//Reset edge detection
IF B1 = 0 THEN
   B1Old := 0;
END_IF;
```

It is easier to copy the code from examples 2 than example 1 to another PLC, because the different PLC types have different one shot standard function blocks.

The execution for examples 2 can be illustrated by this time diagram:

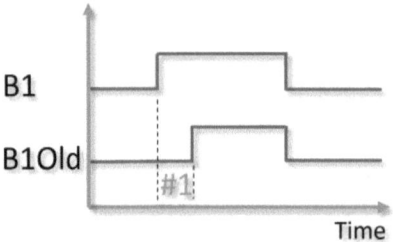

The PLC code #1 is executed immediately after a rising edge on **B1**.

12.2.1 EXAMPEL FB: One Shot rising detection

This chapter shows how the code from example 2, from the previous page, can be moved to a function block, to ensure the code can be reused easily. A function block must be used, because the **CLKOld** variable must be saved after the function call. The code works in the same way as the built-in **R_TRIG** function, therefore the same input and output variable names are used. The block diagram is shown below:

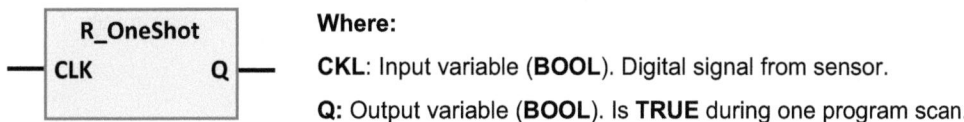

Where:

CKL: Input variable (**BOOL**). Digital signal from sensor.

Q: Output variable (**BOOL**). Is **TRUE** during one program scan.

The syntax "**CLK = 1**" cannot be used in all PLC types. Therefore, this should be changed to either "**CLK = TRUE**" or "**CLK**". To save space the last option is chosen.

Below find the code for the function block and flowchart:

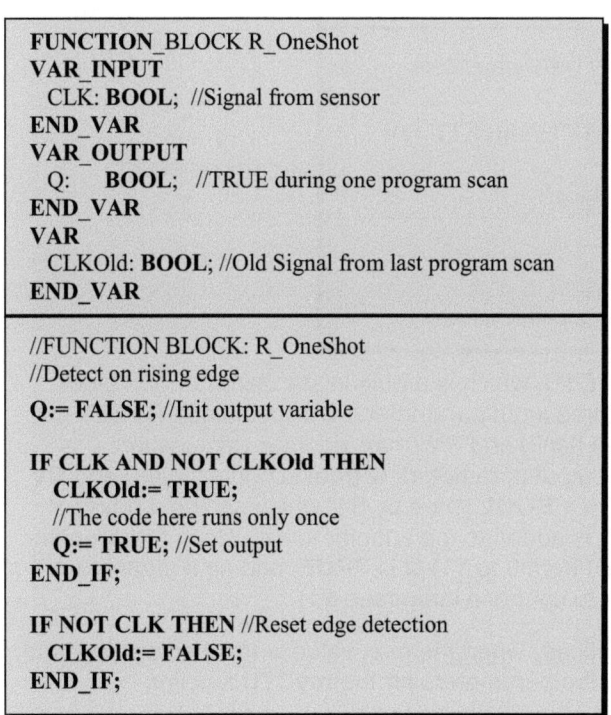

```
FUNCTION_BLOCK R_OneShot
VAR_INPUT
  CLK: BOOL; //Signal from sensor
END_VAR
VAR_OUTPUT
  Q:    BOOL;  //TRUE during one program scan
END_VAR
VAR
  CLKOld: BOOL; //Old Signal from last program scan
END_VAR

//FUNCTION BLOCK: R_OneShot
//Detect on rising edge

Q:= FALSE; //Init output variable

IF CLK AND NOT CLKOld THEN
  CLKOld:= TRUE;
  //The code here runs only once
  Q:= TRUE; //Set output
END_IF;

IF NOT CLK THEN //Reset edge detection
  CLKOld:= FALSE;
END_IF;
```

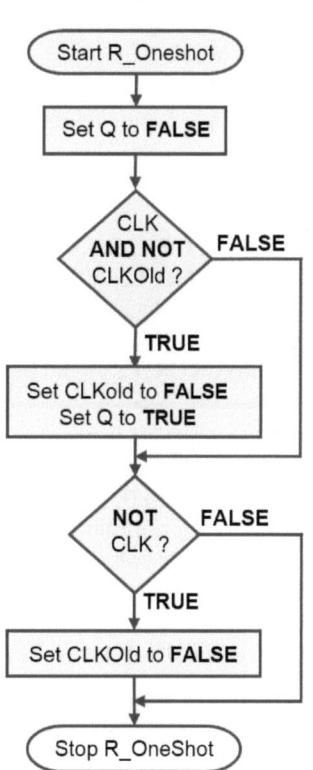

12.3 Counting functions: CTU, CTD, CTUD

A PLC provides three built-in function blocks for counting:

CTU, can count upwards
CTD, can count downwards
CTUD, can count both upwards and downwards.

Below is shown how the **CTU** function block can be used in ST-programming:

```
PROGRAM MAIN
VAR
   myCTU    : CTU;     // Counter UP function
   S1       : BOOL;  // Activate count
   K1       : BOOL;  // TRUE when count is finished
   i        : WORD; // Only for demo and test
END_VAR

// Example 1, counter using the CTU function block
//Counting to 12, auto reset
myCTU (CU:= S1, PV:= 12, RESET:= myCTU.Q);

IF myCTU.Q THEN //Counter done?
   K1 := TRUE;  //#1
END_IF;

i:= myCTU.CV; //Read out current count value
```

A variable **myCTU** is created as a **CTU**, which is a built-in standard function block able to count upwards. **CTU** has three input parameters: **CU** (counting), **RESET** (resetting counter to 0, on the positive flank) and **PV** (max. counter value, where 0 is included in the counting) and two output parameters **Q** (max. counter value) and **CV** (current counter value). **CU** is set to a **BOOL** value by **S1**, which can be a physical switch. Every time it is activated, 1 is added to the counter value. When the counter has reached a value of 12 (counted from 0 to 11) **Q** is **TRUE**, and an **IF** statement sets **K1** to **TRUE**. **K1** can be used to control a lamp (see **#1**).

To make the counter reset automatically when the max value is reached and restarts, **RESET:= myCTU.Q** is inserted in the parameters for the **myCTU** function.

The advantage of the **CTU** function block is that it has **R_TRIG** built-in in the **CU** input. The disadvantage is that it counts internally on a **WORD** variable, and can therefore only count up to 65535. If the **CTU** is used for counting objects on a machine producing an object per minute, an overrun occurs on the internal counter after a period calculated as follows:

60 [objects/hour] => 1440 [objects/day] => 65535/1440 => <u>45,5</u> days.

Below is a solution which can count on a **DWORD** (double **WORD**) variable:

```
PROGRAM MAIN
VAR
  S1_trig : R_TRIG;          // One short
  S1:       BOOL;            // Activate count
  K1:       BOOL := FALSE;   // TRUE when count is finished
  i:        DWORD := 0;      // Counter
END_VAR

// Example 2, Counter with DWORD
S1_trig (CLK:= S1);  // Calling R_TRIG, S1 is input

IF S1_trig.Q THEN  //Count up if positive trig signal
  i:= i + 1;
END_IF;

IF i >= 12 THEN    //Counter done? #1)
  K1:= TRUE;       // Set output
  i:= 0;           // Reset counter
END_IF;
```

```
 R_TRIG
─CLK    Q─
```

K1 becomes **TRUE** when the counter has counted to 12, and at the same time the counter variable **i** is set to 0.
Remark: #1) To create more stable PLC code use "**>=**" instead of only "**=**".

The mode of operation for the two examples (Example 1 vs Example 2) is the same, Example 2 is, however, more usable:

- It can count up to 4.29 billion.

A counter can be used for counting produced parts, amounts of startups on a pump, amount of pulses from instruments: e.g. energy meter or a flowmeter.

12.3.1 EXAMPLE: Counting of items on a conveyor belt

This example is based on code in chapter 12.2.1, page 111, where a solution for a one shot function block was shown.

It is important to test the function block before declaring it complete, because it is difficult to find errors later when the function block is part of a large program.

In order to test the function block, a test program is created as shown below:

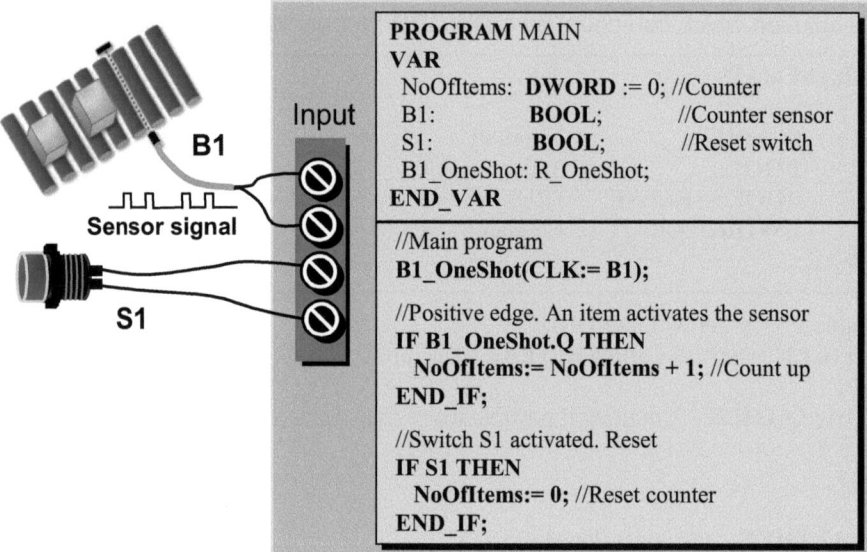

```
PROGRAM MAIN
VAR
   NoOfItems: DWORD := 0; //Counter
   B1:           BOOL;      //Counter sensor
   S1:           BOOL;      //Reset switch
   B1_OneShot: R_OneShot;
END_VAR

//Main program
B1_OneShot(CLK:= B1);

//Positive edge. An item activates the sensor
IF B1_OneShot.Q THEN
   NoOfItems:= NoOfItems + 1; //Count up
END_IF;

//Switch S1 activated. Reset
IF S1 THEN
   NoOfItems:= 0; //Reset counter
END_IF;
```

The **NoOfItems** variable used for counting is defined by a **DWORD** (Double Word) so count can go up to 4.28 million. A **WORD** variable will overflow at 65535.

If the **NoOfItems** variable has to keep its value, when the PLC is turned off, the variable must be declared as **RETAIN** or **PERSISTENT**. See chapter 5, page 28.
B1 is a sensor that detects an item. **B1** is input variable to the **R_OneShot** function block, so only one pulse is detected for each item passing the sensor. When the output variable **Q** is **TRUE**, 1 is added to the **NoOfItems** variable.

S1 is a manual switch to reset the counter variable **NoOfItems** to zero.

The function block is ok, when the **NoOfItems** variable counts 1 up for each item that passes **B1**, and the **S1** switch can reset the **NoOfItems** variable.

12.3.2 EXAMPLE FC: Instrument pulse counter

Dette This section shows a function block which can be used to collect pulse signals from measuring instruments. Some instruments have a digital output that sends off a pulse each time the instrument has measured a certain quantity. It can be a Watt Meter (power meter) that measures energy consumption and gives a pulse every time it has measured 0.1 KWh. Or a flow switch that gives a pulse every time it has measured 100 liters.

The code is a rework of the code shown in chapter 12.2.1 page 111.
The code is changed to a function block, so it can be used for multiple instruments. The counter value is now an input parameter to the function block, so the function block can easily be used for several different types of instruments:

```
FUNCTION_BLOCK PulseCount
VAR
 Pulse_OneShot: R_OneShot; //R_TRIG;
END_VAR
VAR_INPUT
 Pulse:    BOOL; //Pulse from Instrument
 Amount: REAL := 0; //Add Pulse amount
 Reset:   BOOL := FALSE; //Reste counter
END_VAR
VAR_OUTPUT
 AmountTotal: REAL := 0; //Read out
END_VAR
```

```
//PulseCount FUNCTION BLOCK
Pulse_OneShot(CLK:= Pulse);

//Positive edge trig puls signal from instrument
IF Pulse_OneShot.Q THEN
   AmountTotal:= AmountTotal + Amount; //Count up
END_IF;

//Reset counter. Manuel or if very big value
IF Reset OR  AmountTotal > 100000000 THEN
   AmountTotal:= 0;
END_IF;
```

Below find an example for testing the function block:

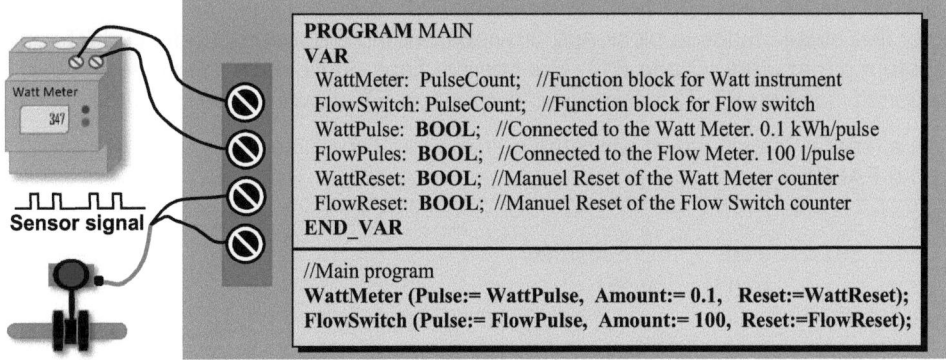

```
PROGRAM MAIN
VAR
   WattMeter: PulseCount;  //Function block for Watt instrument
   FlowSwitch: PulseCount;  //Function block for Flow switch
   WattPulse: BOOL;   //Connected to the Watt Meter. 0.1 kWh/pulse
   FlowPules: BOOL;   //Connected to the Flow Meter. 100 l/pulse
   WattReset: BOOL;   //Manuel Reset of the Watt Meter counter
   FlowReset: BOOL;   //Manuel Reset of the Flow Switch counter
END_VAR
```

```
//Main program
WattMeter (Pulse:= WattPulse, Amount:= 0.1,  Reset:=WattReset);
FlowSwitch (Pulse:= FlowPulse, Amount:= 100, Reset:=FlowReset);
```

AmountTotal contains the total amount measured and can be reset by **Reset**.

Note that high speed input counter cards or input modules often have to be used, as the digital sensor signals from instruments often come in high speed.

12.4 Repeated program 'calls' and timer delay: TON, TOF

In a PLC program, some equipment must only be turned on for a certain period of time. For example, a motor could be programmed to run for 30 minutes per hour, the light in a staircase programmed to switch off automatically after a period of time or a stop watch. An example could also be an alarm signal from a level sensor in a tank which should not go off until after a certain period of time, because wave motions in the tank can affect the level sensor measurements. A timer solves these problems.

There are two types of standard timers in a PLC:

TON (On-delay timer, ODT, TONR, ON delay) *Delayed connection relay*

A **TON** timer function block sets a **BOOL** variable **Q** to **TRUE** after a certain period of time indicated by **PT**.
Can be used if a component must receive a signal after a certain period of time in order to start.
Used for Noise Attenuation in an ON/OFF switch.
The time where **IN** is activated must be longer than **PT**.

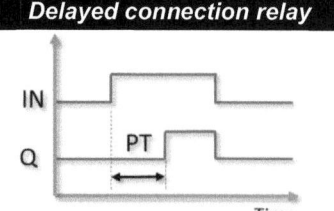

TOF (Off-delay timer, OFFDT, TOFR, OFF delay) *Delayed detection relay*

A **TOF** timer function block sets a **BOOL** variable **Q** to **FALSE** after a certain period of time indicated by **PT**.

Can be used for light in a staircase or toilet ventilation, where the system must be powered off after a period of time. The time starts after **IN** is set to **FALSE**.

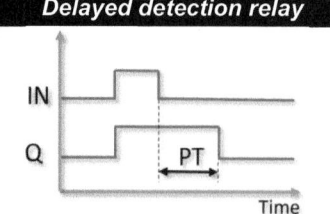

A timer is a built-in function block and provides two input parameters (**IN** and **PT**) and two output parameters (**Q** and **ET**). The positive flank on **IN** starts the timer and the time period is set on **PT**. **Q** is the signal output and **ET** shows the current time.

Below a timer is shown which will remain active for 100 milliseconds after **S1** has become **FALSE**.

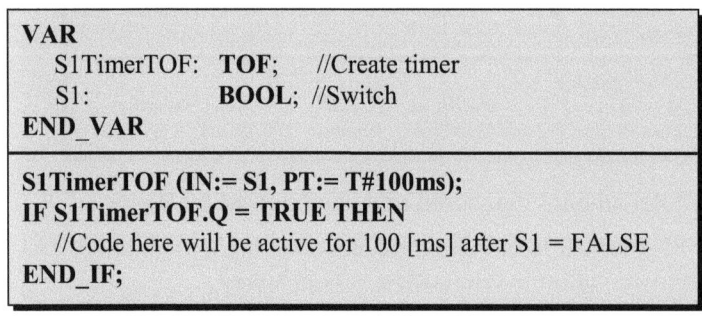

```
VAR
    S1TimerTOF:  TOF;     //Create timer
    S1:          BOOL;  //Switch
END_VAR

S1TimerTOF (IN:= S1, PT:= T#100ms);
IF S1TimerTOF.Q = TRUE THEN
    //Code here will be active for 100 [ms] after S1 = FALSE
END_IF;
```

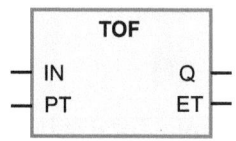

Example 2 below, shows how a timer can be implemented including an automatic restart. The timer is active for 10 seconds and restarts automatically.

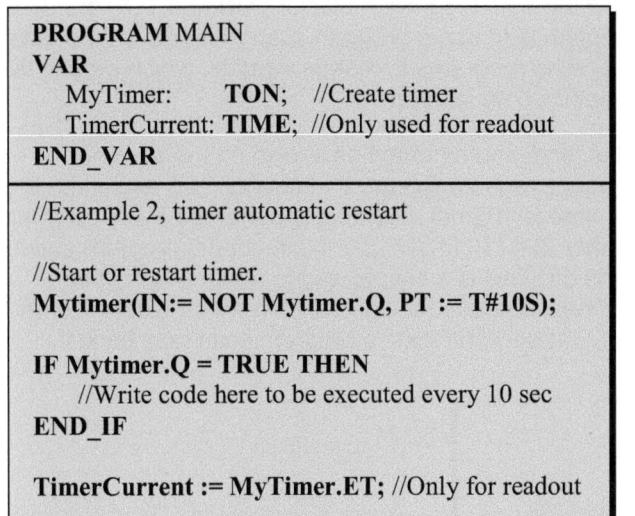

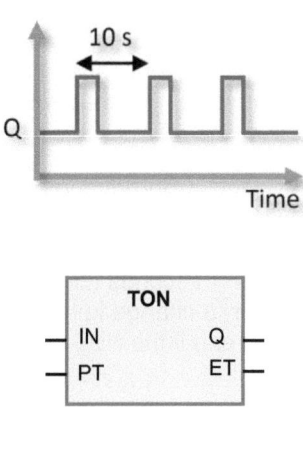

```
PROGRAM MAIN
VAR
    MyTimer:      TON;   //Create timer
    TimerCurrent: TIME;  //Only used for readout
END_VAR

//Example 2, timer automatic restart

//Start or restart timer.
Mytimer(IN:= NOT Mytimer.Q, PT := T#10S);

IF Mytimer.Q = TRUE THEN
        //Write code here to be executed every 10 sec
END_IF

TimerCurrent := MyTimer.ET; //Only for readout
```

The mode of operation is as follows:

MyTimer: The data type is a **TON** function block..

TimerCurrent: Only used to be able to read the current value of the timer – an efficient tool to make it all work.

The current value on the timer is read on the last line in the PLC code, and is read out by copying the value, indicated on **MyTimer.ET**, to **TimerCurrent**, which is created with the data type TIME, because it is of the same data type as **MyTimer.ET**.

When the timer is active and running **MyTimer.Q** = **FALSE**. When the timer has expired, **MyTimer.Q** = **TRUE** and the timer stops. The timer restarts automatically, because **IN** is the inverted value of **MyTimer.Q** (use **NOT** before **MyTimer.Q**).
The parameter **PT** sets up the time delay, where time is indicated by **T#** and a digit (in this example 10) followed by the SI-unit (s = second, ms = millisecond, h = hour).

If the timers are to run very fast, select **LTON** or **LTOF** function block.

Using program-scan as a timer
Another way of implementing a timer is by using the PLC program-scan time.
Read about this in the next chapter.

12.4.1 EXAMPLE: Using the program scan as timer

The previous pages describe how the built-in function blocks **TON** and **TOF** are used to create a time delay. Another option is to use a program scan to create a time delay. This is possible because the PLC runs programs in real-time mode, which means that programs are executed with a specific time interval.

Below are two program examples, both implemented as shown on the flowchart below. The program contains a count variable **Count,** that counts 1 up each time the program is executed. If the scan time is 10 [ms], counting is perfomed every 10 [ms]. This causes the light to turn on after 200 (10 [ms] * 200 = 2 seconds) program scans, and after 4 seconds the light turns off. This is a simple way to make a light flash. However, a change in the scan time will cause the light to flash with a different time interval, and if the PLC is heavily loaded with tasks, a program scan may be lost.

```
PROGRAM LampFlash
VAR
    Count: INT := 0;   //Counter value
    Lamp: BOOL;        //Connection to the Lamp
END_VAR
```

```
//Program example 1 (scan time 10 ms)
Count:= Count + 1;  //Count up

IF Count > 200 THEN
    Lamp:= TRUE;  //Light on
ELSE
    Lamp:= FALSE;  //Light off
END_IF;

IF Count > 400 THEN
    Count:= 0; //Reset counter
END_IF;
```

```
//Program example 2 (Scan time 10 ms)
Count:= count + 1;  //Count up

Lamp:= Count > 200;

IF Count > 400 THEN
    Count:= 0;  //Reset counter
END_IF;
```

Start
LampFlash

Count up

Count >200 ? — **TRUE** → Turn Lamp on

FALSE

Turn Lamp off

Count > 400 ? — **TRUE** → Reset counter

FALSE

Stop LampFlash

12.4.2 EXAMPLE: Function block for Flashing Light

Many machines have a Light Tower to inform the operator about the operating state of the machine. If one of the lights in the tower needs to flash, the code must be incorporated in the PLC program. Below find a function block that can be used to make lights flash.

The function block uses two **TON** timers. The **TON** timer defines the time the light is turned off and on. When one timer ends, the other timer starts. The input variable **Value** sets the period of time the light must be off and on for.

```
FUNCTION_BLOCK FB_LightFlash
VAR
  TimerOn: TON;
  TimerOff: TON;
END_VAR
VAR_INPUT
  Enable:  BOOL;  //Running code if TRUE
  Value:   TIME;  // Value for timers
END_VAR
VAR_OUTPUT
  Q:      BOOL;  // Return Value
END_VAR
```

```
//FUNCTION BLOCK for a Flashing light

Q := FALSE; //Set FALSE if not enabled

IF Enable THEN
  TimerOn(IN:= TimerOff.Q, PT:= Value);
  TimerOff(IN:= NOT TimerOn.Q, PT:=Value, Q=>Q);
END_IF;
```

Below find a solution to test the function block. When the **B1** switch is activated, the red light flashes at one second intervals. When the **B2** switch is activated the yellow light flashes at two second intervals. When the **B3** is activated, the green light remains on.

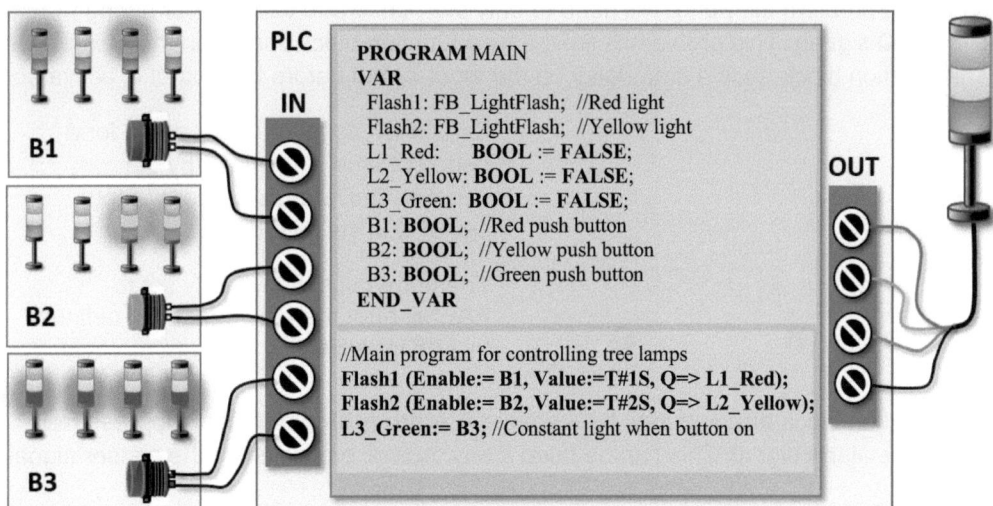

12.4.3 EXAMPLE FC: Time delay on digital alarms

In many control solutions, it is often required to have a time delay on digital sensor signals because contact noise from the sensors can result in many on/off signals. If the sensor is used directly to control a motor, the motor will start and stop many times, which can damage the motor.

The example below is of a flow switch installed in a pump well. When the pump well level rises during inflow of water, the flow switch will be activated (**IN** signal). However, the water inflow can result in waves which can result in many on/off signals as illustrated by (1):

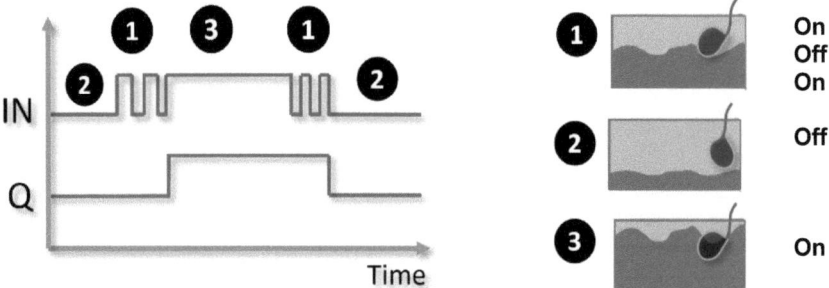

If the pump in the well was directly controlled by the **IN** signal, the pump will start and stop many times before reaching a state where the pump will run continuously (3). To solve this problem the pump will need to only start once and then run for a long period of time (**Q** signal). This problem is solved by a function block shown on the next page. The function block adds a time delay on the input signal before the output is activated:

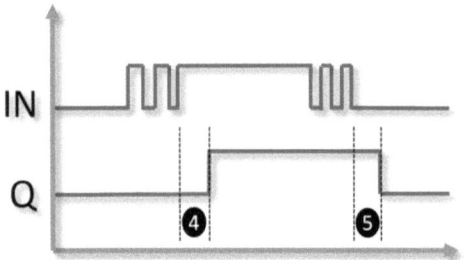

Sensor **IN** signal must be stable for a period of time (4) defined by the variable **OnDelaySec** before output signal **Q** is set to **TRUE**.

The output signal **Q** is only set to **FALSE** after the time period (5) defined by **OffDelaySec**

The function block also includes the variable **AlarmInhibit** which can suppress (not show) the alarm signal. This can be used if any fault or error arise in the sensor signal.

Program code suggestions:

```
FUNCTION_BLOCK
AlarmOnOffDelay
VAR_INPUT
  IN: BOOL := FALSE;
    //Delay on time. Disabled when 0
  OnDelaySec: WORD := 0;
    //Delay off time. Disabled when 0
  OffDelaySec: WORD := 0;
    //Q always off
  AlarmInhibit: BOOL := FALSE;
END_VAR
VAR_OUTPUT
  Q: BOOL;  //Out signal
END_VAR
VAR
  TimerOn: TON; //Internal timer
  TimerOff: TOF;  // Internal timer
END_VAR
```

```
//Code for FUNCTION BLOCK: AlarmOnOffDelay
//On delay timer
TimerOn(IN:=IN, PT:=WORD_TO_TIME(OnDelaySec*1000));

//Set alarm Out
Q:= TimerOn.Q;

//Special case when OnDelayAlarm is zero (0)
IF (OnDelaySec = 0 AND IN) THEN
  Q:= TRUE;
END_IF;

//Off Delay timer
TimerOff(IN:=IN,PT:=WORD_TO_TIME(OffDelaySec*1000));

//Set Alarm out
IF TimerOff.Q AND NOT IN THEN
  Q:= TRUE;
END_IF;

//Special case when OnDelayAlarm is zero (0)
IF (OffDelaySec = 0 AND NOT IN) THEN
  Q:= FALSE;
END_IF;

//If Alarm Inhibit. Must be the last line
Q:= Q AND NOT AlarmInhibit;
```

An example using the function block is shown below:

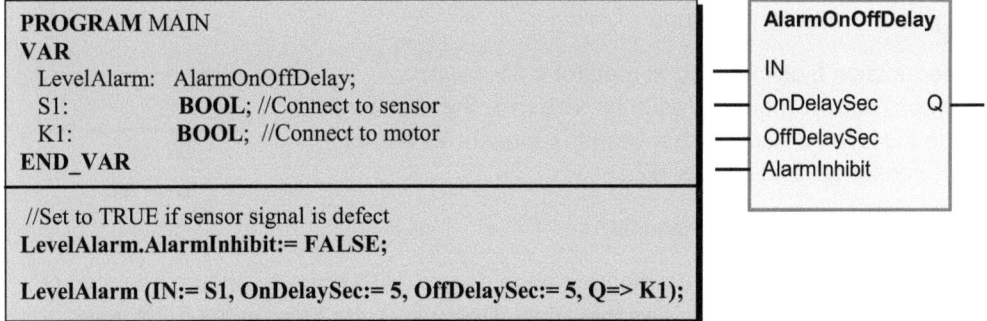

```
PROGRAM MAIN
VAR
  LevelAlarm:   AlarmOnOffDelay;
  S1:           BOOL; //Connect to sensor
  K1:           BOOL; //Connect to motor
END_VAR

//Set to TRUE if sensor signal is defect
LevelAlarm.AlarmInhibit:= FALSE;

LevelAlarm (IN:= S1, OnDelaySec:= 5, OffDelaySec:= 5, Q=> K1);
```

The time delay is defined by the variables **OnDelaySec** and **OffDelaySec**. The time is in seconds and the data type is **WORD**. The **WORD_TO_TIME** function converts the value into the time format the **TON** and **TOF** function requires. The value is multiplied by 1000, because the time format must be in [ms].

12.4.4 EXAMPLE FC: Monitoring of analog values and alarms

This example describes a function block that can be used to monitor an analog sensor value inside or outside a specified range. If the value is outside the range, an alarm will be triggered. The alarm should only be triggered when the sensor value has remained outside of the range for a certain period of time. The time delay prevents the alarm being triggered if the value only momentarily goes outside the range.

This function block can e.g. be used to monitor:

- A temperature range, to turn on heating at low temperatures.
- Large pressure differences on a filter, to alert that the filter must be cleaned or changed.
- The power consumption of a machine. If it is very high its causes may need to be investigated

The function block is designed to be used for several purposes. The measurement range is defined by an upper limit (2) **LimitH** and lower limit (3) **LimitL**

If the value measured by an analog sensor (1) is outside the limit range for a long period of time (5), **Q** will be **TRUE**.

If the value is outside the boundary range for only a short period of time (4), **Q** will not be **TRUE**.

An alarm (6) is triggered as soon as the value is outside the range

If the function block is used to monitor a filter with a pressure sensor mounted each on side of it, the measured value (1) is the difference between the measured values from sensor **B1** and sensor **B2**.

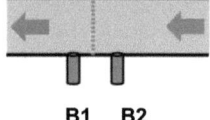

The filter may be for air or liquids.

The function block shown on the next page uses a **TON** timer to determine when **Q** should be **TRUE**. The input variable **AlarmDelay** sets the time on the **TON** timer.

In addition, the function block swaps the limit values **LimitH** and **LimitL** if the high limit is the lowest limit. This prevents programming errors.

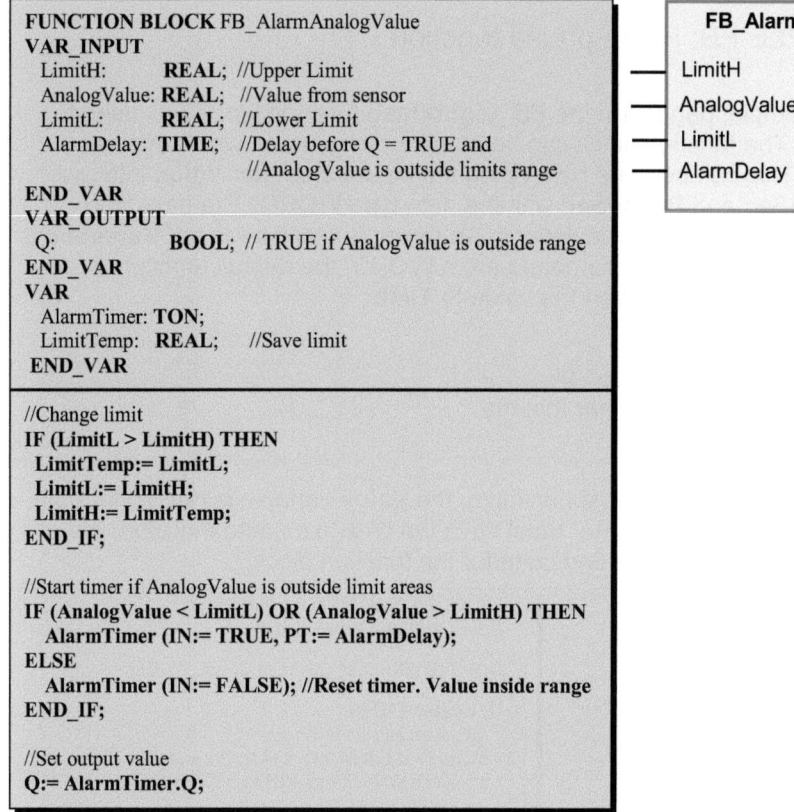

```
FUNCTION BLOCK FB_AlarmAnalogValue
VAR_INPUT
  LimitH:        REAL;  //Upper Limit
  AnalogValue: REAL;   //Value from sensor
  LimitL:        REAL;  //Lower Limit
  AlarmDelay: TIME;   //Delay before Q = TRUE and
                       //AnalogValue is outside limits range
END_VAR
VAR_OUTPUT
  Q:              BOOL;  // TRUE if AnalogValue is outside range
END_VAR
VAR
  AlarmTimer: TON;
  LimitTemp:  REAL;     //Save limit
END_VAR
```

```
//Change limit
IF (LimitL > LimitH) THEN
  LimitTemp:= LimitL;
  LimitL:= LimitH;
  LimitH:= LimitTemp;
END_IF;

//Start timer if AnalogValue is outside limit areas
IF (AnalogValue < LimitL) OR (AnalogValue > LimitH) THEN
  AlarmTimer (IN:= TRUE, PT:= AlarmDelay);
ELSE
  AlarmTimer (IN:= FALSE); //Reset timer. Value inside range
END_IF;

//Set output value
Q:= AlarmTimer.Q;
```

Program for testing the function block:

```
PROGRAM MAIN
VAR
  MeasVal:     REAL;  //Sensor value
  AlarmQ:      BOOL;  //Alarm from the FUNCTION BLOCK
  AlarmCheck: FB_AlarmAnalogValue;
END_VAR
```

```
//Test code
MeasVal:= 8; //If MeasVal = 2 the AlarmQ variable will be TRUE after 20 seconds
AlarmCheck (LimitH:=10, AnalogValue:= MeasVal, LimitL:= 4, AlarmDelay:= T#20S);
AlarmQ:= MeasureAlarm.Q;
```

12.4.5 EXAMPLE FB: Pulse pause function

In the previous example, page 119, the **FB_LightFlash** function block was used to make a light flash. The function block can be made more general, which ensures more code reuse. This can be done by dividing the input parameter **Value** into two parameters: **PulseSec** and **PauseSec** with the data type **WORD**. The data type WORD makes it easier to make calculations. An example could be to set **PulseSec** to 25% of **PauseSec**. When input parameters are a WORD, the default function **WORD_TO_TIME** is used to convert the value to **TIME**.

Timeline for the function block:

The variable name has been chosen by adding **Sec** to the name, so it is clear that the unit is seconds.

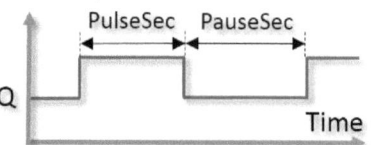

To keep the number of variables to a minimum, the **Value** variable is reused for both **TimerOn** and **TimerOff**. A **TON** timer must have the time in ms and therefore the value is multiplied by 1000. Below find code for the function block:

```
FUNCTION_BLOCK FB_PulsePause
VAR
  TimerOn: TON;
  TimerOff: TON;
  Value:     TIME; //Internal conversion value
END_VAR
VAR_INPUT
  Enable:   BOOL;  // Run code if TRUE
  PulseSec: WORD;  // Value for on signal
  PauseSec: WORD;  // Value for off signal
END_VAR
VAR_OUTPUT
  Q: BOOL; // Return Value
END_VAR
```

```
//FUNCTION BLOCK for PulsePause
//Used for Pulse Pause signal

Q := FALSE; //Set to FALSE if not enabled

IF Enable THEN
  //Convert and set Off timer
  Value:= WORD_TO_TIME (PauseSec * 1000);
  TimerOn(IN:= TimerOff.Q, PT:= Value);

  //Convert and set On timer
  Value:= WORD_TO_TIME (PulseSec * 1000);
  TimerOff(IN:= NOT TimerOn.Q, PT:=Value, Q=>Q);
END_IF;
```

Example of using the function block (as an alternative to chapter 12.4.2, page 119):

```
Flash1 (Enable:= B1, PulseSec:= 1, PauseSec:=1, Q=> L1_Red);
```

The function block can e.g. be used for administering a chemical liquid where the valve is controlled by Q. **PauseSec** is the time the chemical needs to dissolve. By changing **PulseSec** the administered amount can easily be changed.

The function block can also be used for **P**ulse **W**ide **M**odulation (PWM) or pulse train which require a digital output card to be of the high speed type.

12.4.6 EXAMPLE FB: A timer with a pause function

There may be a need to pause a timer for a period of time. With the function block **TONP** below, it is possible to pause a **TON** timer. The function block works in the same way as the **TON** function block, but **PAUSE** is added as an input parameter. This means that the timer is put on hold when **PAUSE** = **TRUE** and continues when **PAUSE** = **FALSE**.

```
FUNCTION_BLOCK TONP
VAR_INPUT
  IN :        BOOL;   // Start timer. Must be true when timer is running
  PT :        TIME;   // Set time. In the format like T#1S, T#10ms
  PAUSE :     BOOL;   // Timer paused when TRUE
END_VAR
VAR_OUTPUT
  Q :         BOOL;   // Timer ended. TRUE when time ended
  ET :        TIME;   // Current time
END_VAR
VAR
  PauseOld :  BOOL := FALSE; // Handle Oneshot
  TimerPause : TIME;  // Time when paused
  Timer :     TON;   // Internal timer
END_VAR
```

```
//The FUNCTION block can be used to pause a TON timer

//Reset timer
IF NOT IN THEN
  TimerPause := T#0S;
END_IF

//Stop timer when input parameter PAUSE changes from FALSE to TRUE
IF PAUSE AND NOT PauseOLD THEN
  PauseOld:= TRUE;
  TimerPause := TimerPause + Timer.ET; // Save current time
END_IF

Timer (IN := IN AND NOT PAUSE, PT := PT - TimerPause);

//Reset pause TRIG signal
IF NOT PAUSE THEN
  Pauseold:= FALSE;
END_IF;

//Set output values
Q := Timer.Q;
ET := TimerPause + Timer.ET;
```

TONP

IN	Q
PT	ET
PAUSE	

13   Special functions and program structures

This chapter describes a number of special functions, commonly used program structures and more complex programs.

13.1   Simple queue structure

This example describes the simplest implementation of a queue. A queue is used when e.g. there are many packages on a conveyer belt, waiting for treatment by a machine in a large plant. The packages often require information like weight, receiver, size or content. Weight gives information about a package which must be saved in a queue, so that the information can follow the package through the plant. If the package has a readable bar code, it is not necessary to implement a queue, as the information about the package can be accessed from a shared database – i.e. the company's production control system, often named:

> **M**anufacturing **E**xecution **S**ystems (**MES**),
> **M**anufacturing **I**nformation **S**ystems (**MIS**) or
> **W**arehouse **C**ontrol **S**ystem (**WCS**)

When implementing a queue, the objects must not change their place in the queue. However, if the packages provide e.g. a bar code or any kind of ID the packages may change their place in the queue.

An **ARRAY** should be created with the maximum length the queue is expected to be. Exceeding the required length of the **ARRAY** will take up memory unnecessarily and increase the execution time of the program.

A simple example is shown below where an **ARRAY** with 6 positions of the data type **INT** is created. Firstly, all the **ARRAY** positions are initialized to -1, because -1 can be used to check whether the position is empty:

```
PROGRAM MAIN
VAR
   Que: ARRAY[QueMin..QueMax] OF INT;
   n: INT; //Counter to FOR loop
END_VAR
VAR CONSTANT
   QueMax: INT := 5;
   QueMin: INT := 0;
END_VAR

FOR n:= QueMin TO QueMax DO
   Que[n]:= -1; //Init ARRAY
END_FOR;
```

The number above the **ARRAY** shows the position no:

0	1	2	3	4	5
-1	-1	-1	-1	-1	-1

Next: The **ARRAY** is now filled with three values (23, 35, 71). Values are inserted into the **ARRAY** from left to right, so that the value inserted first (with the value 23), is positioned all the way to the left, and the value inserted last is positioned all the way to the right on position 2 (with the value 71) as shown below

0	1	2	3	4	5
23	35	71	-1	-1	-1

Inserting values in the queue can be carried out with this PLC code, where the **ARRAY** is named **Que:**

```
Que [0] := 23;
Que [1] := 35;
Que [2] := 71;
```

The oldest value in the queue is 23, and is also the value which is taken out first. The simplest way to keep control of the queue, is to make sure that the oldest value is always positioned at position 0.

When the oldest value is taken out, all the values have to be moved one position to the left. The next value to be taken out is therefore 35.

A **FOR** loop is used to move all the values one position to the left. The values are always moved left to not overwrite the values which already exist in the queue. The **FOR** loop must be executed one time less than the maximum positions in the **ARRAY** (array length), as shown in the example below:

FOR n:= 0 **TO** 5 - 1 **DO**
    Que [n]:= Que [n + 1];
END_FOR;

The **ARRAY** has 6 positions and the values have to be moved 5 times. Therefore, the **FOR** loop is executed 5 times as illustrated below:

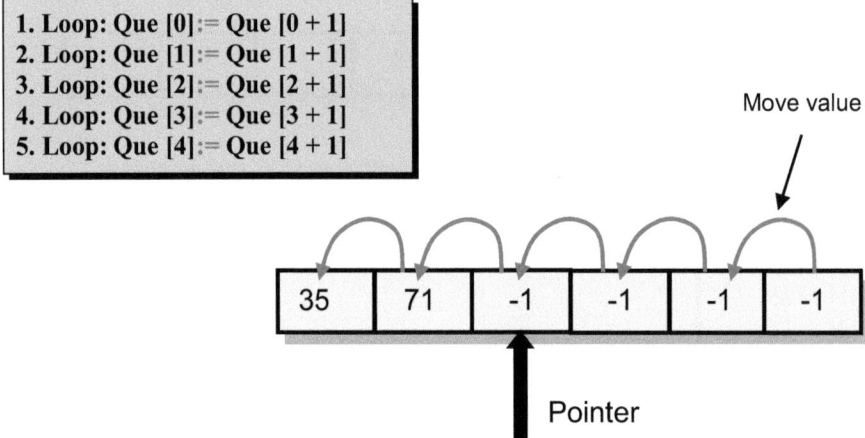

1. Loop: Que [0]:= Que [0 + 1]
2. Loop: Que [1]:= Que [1 + 1]
3. Loop: Que [2]:= Que [2 + 1]
4. Loop: Que [3]:= Que [3 + 1]
5. Loop: Que [4]:= Que [4 + 1]

Move value

| 35 | 71 | -1 | -1 | -1 | -1 |

Pointer

To keep track of which position the next value must be inserted at, a variable must be used called index or pointer. It starts by pointing at position 0, because the queue is empty. Every time a new value is inserted into the queue, the pointer is moved one position to the right, and if a value is removed (taken out) from the queue, the pointer have to be moved one position left.

The disadvantage of the simple queue is that a lot of time is spend executing the whole queue to 'move'/'push' the values every time a value is taken out. To solve this problem a circular buffer can be used which uses pointers instead of moving the values, every time a value is taken out or inserted.

A queue is often called **FIFO** – meaning **F**irst **I**n **F**irst **O**ut. The value which is inserted as the first value has to be taken out first. This is described in the next chapter.

13.2 FIF0 – First In First Out

The previous chapter described the implementation of a simple queue, where all values are moved every time a value is removed from the queue. This chapter describes a queue, where the values ARE NOT MOVED, when a value is taken out. This makes the PLC code efficient.

An efficient **FIFO** consists of an array and two pointers. The pointers points to a position in the array, as shown in the below illustration:

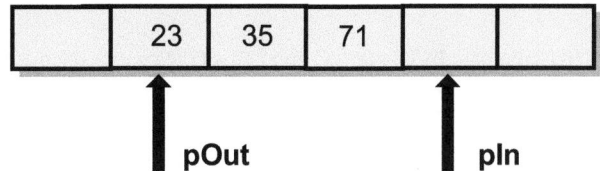

The pointer **pOut** is pointing at the value which must come out of the queue first, and the pointer **pIn** is pointing at the next free position in the queue. Every time a value is removed from the queue, the pointer **pOut** is moved one position to the right. Every time a new value is inserted where the pointer **pIn** is pointing, the pointer **pIn** is moved one position to the right. When a pointer comes to the end of the array, it will be moved to the beginning of the array.

   A **FIFO** is also called a circular buffer.

The different PLC types usually offer a **FIFO** in the built-in software library. It can be used, but it is often not possible to make changes to it to meet your requirements (it is often locked with a manufacturer password). Furthermore, using the built-in **FIFO** code can limit the possibilities to transfer the program to another PLC type. The next pages cover an implementation of a **FIFO**. The code can be used for learning purposes, and feel free to adjust the code to meet your requirements.

To limit the number of variables going into the function block, a control variable, named **INOutStatus** is used. This variable can have three status settings:

   0   Do nothing inside the function block
   1   Insert the value from **DataIn** into the queue
   2   Take out a value from the queue and place it in the **DataOut** variable.

Two **BOOL** variables could be used instead of the **InOutStatus** control variable, because **BOOL** variables would be easier to work with in a Ladder Diagram program.

Below find the variables for the function block:

```
FUNCTION_BLOCK FIFO
VAR_INPUT
    DataIn :        REAL;  // Insert data into buffer
    INOutStatus :   INT;    // 0 : Do nothing, 1 : Insert data, 2 : Take out data
END_VAR
VAR_OUTPUT
    DataOut :       REAL;  // Take out data from the buffer
END_VAR
VAR CONSTANT
    BufferMax :     INT := 5;  // Max fixed size of the buffer
    BufferMin:      INT := 1;  // Min fixed size of the buffer
END_VAR
VAR
    NoOfDataPoints : INT := 0; // Current no. of data points (elements)
                    // Array including all elements
    Buffer:         ARRAY[BufferMin..BufferMax] OF REAL;
    pIn :           INT := 1; //Pointer to first element
    pOut :          INT := 1; //Pointer to last element
END_VAR
```

Below find a demo program using the function block. The demo code shows two values being inserted and one value taken out from the FIFO.

The code below must only be executed once. Use **S1** for executing once.

```
PROGRAM MAIN
VAR
    OutData:   REAL;      //Value from the queue
    MyFIFO:    FIFO;
    S1_Trig :  R_TRIG;  //One shot for S1
    S1         BOOL:= FALSE; //Test button
END_VAR

//Test of the FIFO
S1_Trig (CLK:= S1);
IF S1_Trig.Q THEN
    //Insert 71 and 35 into the FIFO
    MyFIFO (DataIn:= 71, INOutStatus:= 1);
    MyFIFO (DataIn:= 35, INOutStatus:= 1);

    //Take out the first inserted value,  OutData = 71
    MyFIFO (INOutStatus := 2 , DataOut => OutData);
END_IF;
```

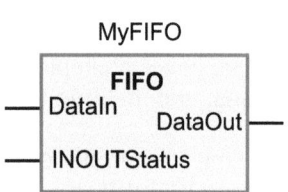

MyFIFO

```
///////////////////////////////////////////////////////////////////////////////////////////
// FIFO - First In First out
// Can handle up to BufferMax REAL data points
// If more REAL data points are entered, the old ones will be overwritten
///////////////////////////////////////////////////////////////////////////////////////////

//Insert data into buffer
IF INOutStatus = 1 THEN
  IF pIn <= BufferMax THEN
    Buffer[pIn] := DataIn; //Insert
    //Increase number of data points
    IF NoOfDataPoints < BufferMax THEN
      NoOfDataPoints:= NoOfDataPoints + 1;
    END_IF
    pIn:= pIn + 1; //Set to next element
  ELSE // buffer full, insert into first element
    pIn:= BufferMin;
    Buffer[pIn] := DataIn;
    //Move pointer to next element
    pIn:= pIn + 1;
  END_IF;
END_IF;

//Take out data from the buffer
IF INOutStatus = 2 THEN
  IF NoOfDataPoints > 0 THEN //There must be data
    Dataout:= Buffer[pOut];
    Buffer[pOut] := 0; //Set to 0 to show that the value is removed
    NoOfDataPoints:= NoOfDataPoints - 1;
    IF pOut < BufferMax THEN
      pOut:= pOut + 1;
    ELSE
      pOut:= BufferMin;
    END_IF;
  END_IF;
END_IF;

//Is buffer full? Last value is overwritten, move pIn pointer
IF NoOfDataPoints >= BufferMax THEN
  pIn := pOut;
END_IF;
```

13.3 Generating random numbers (RND, Randomize)

This chapter shows how a few lines of code can generate random numbers. The random numbers can be used for testing a PLC program, where numbers can e.g. be the weight or the size of parts which must be packed in boxes. In this way, the PCL program can be tested in the office with many different numbers – a test which is very close to a test carried out with real parts.

Often no access is given to real production of parts to test the PLC program, so by simulating the parts with a random number generator, as shown below, it is possible to test a large amount of PLC code in the office, before the commission test.

By testing the PLC code in the early phases of development, possible programming faults and bugs are found and corrected. They are always more difficult to find later.

The PLC code is written in a function block named **RND**:

```
FUNCTION_BLOCK RND
VAR_INPUT
   Seed:          INT;  // Start value, a value below ValueMax
   ValueMax:      INT;  // Max value to be generated
END_VAR
VAR_OUTPUT
   ValueRandom:  INT;  // The returned randomized value
END_VAR
VAR
   RandomSeed:  DINT := 0;  //Start value.
END_VAR
```

When running the program, the first time around, the **ValueMax** must be set to the maximum number to be generated. If **ValueMax** is set to 12, the return value **NewValue** will be a random number between -12 and 12 after each program execution. When all numbers between -12 and 12 have been 'drawn', the process is repeated from the beginning. Note that the numbers occur in the same order and the distribution is mathematically evenly distributed throughout the whole interval -12 to 12.

Having the same start value for **Seed**, numbers appear in the same sequence. **Seed** can be taken from the built-in clock in the PLC to ensure different start values and increased number randomization.

The function block can only return integer numbers. If decimal numbers are required, the **ValueMax** variable can be multiplied by 10. Then the **NewValue** has to be divided by 10 to get a random decimal number.

```
///////////////////////////////////////////////////////////////////////////////////////////////////////////
// This function is a randomize number function
//
// The function generates a different number each time the function is called
// The seed value set the start value and this can be taken from the PLC
// main clock time to ensure different start numbers
// Refer to: "The C Programming Language," by Kernighan and Ritchie:
//
// INPUT: Valuemax is the max value ( + / - ) of the range
// INPUT: Seed, starts one number below max
// OUTPUT: ValueRandom a number in the range - ValueMin and ValueMax
IF RandomSeed = 0 THEN //Init
   RandomSeed := Seed;
END_IF
RandomSeed :=  RandomSeed * 1103515245 + 12345;
ValueRandom := DINT_TO_INT((RandomSeed / 65536) MOD (ValueMax + 1));
```

The **RND** function block can be tested by the following:

The variable **MyRND** is created in the **MAIN** program, with the data type **RND** and a variable **NewValue** is created to contain the random number.

In the demo code below the **ValueMax** is set to 12 because it is the maximum number when using two dice. The **RND** function block can return zero (0) which cannot be used when we are using two dice. And because the **RND** function block return values between -12 and 12 the **ABS** function is used to only have a positive value.

```
PROGRAM MAIN
VAR
   MyRND:   RND;
   NewValue: INT;  //New random value
   Dice :     INT;  //The two dice played
   Mytimer:  TON;  //Timer to have a delay between each play
END_VAR

Mytimer(IN:= NOT Mytimer.Q, PT:=T#5S); //Auto reset timer

IF Mytimer.Q = TRUE THEN // Play the dice when the time is up
   MyRND(Seed:= 5, ValueMax:=12, ValueRandom => NewValue);

   IF NewValue <> 0 THEN //We don't like zero
      Dice:= ABS (NewValue); //Always use a positive value
   END_IF;
END_IF;
```

13.4 Digital low-pass filter (LP-filter)

This chapter describes the implementation of a digital low-pass filter. This filter is based on a **L**ow **P**ass (LP-Filter), consisting of an electronic coil in serial connection with an electronic capacitor (RC-filter). This filter lets the low frequencies pass through and remove the high frequencies and can be used to remove noise signals. On the analogue input module, an LP filter is normally built-in where it is possible to filter noise and unwanted deflections from sensors and measuring equipment. Normally, it is not possible to modify the filter frequency online on an analogue input module. In some plants and machinery an online change of the filter frequency is required, and to perform this a digital filter in the PLC program is needed.

The example shown below is a 1st order digital filter. Also called an exponential filter.

A Fourier transformation (advanced math) is used for transferring the analogue filter to a digital filter.

There are types of filters for Digital Signal Processing (DSP) on the market, and the FIR (Finite Impulse Response) is among the most well-known. The advantage of using a digital filter instead of an average of data such as e.g. 'moving average' is that a 'moving average' includes all values and uses a long **ARRAY** to contain these. A digital filter removes the outlier values and is fast for a PLC to work with.
A function block is used for the implementation, because the filter must use a value from the previous program scan and this value is saved in **ValueOld**.

The filter frequency is adjusted by modifying the filter constant k:

k	Curve	Description
> 0.01	2	The filter is fast and does not remove a lot of signal.
1	1	The filter is not working (filter turned off).
< 0.01	3	A lot of signal is filtered out (cutoff), and the signal takes a long time to come into the right signal level.

```
FUNCTION_BLOCK LP_Filter
VAR_INPUT
    ValueRaw :      REAL;   // Input value
    k :             REAL;   // Filter constant
END_VAR
VAR_OUTPUT
    ValueFiltered : REAL;   // The filtered output value
END_VAR
VAR
    ValueOld :      REAL;   // Value from last scan
END_VAR


/////////////////////////////////////////////////////////////////////////
//First-order lag filter (LP-Filter)
/////////////////////////////////////////////////////////////////////////
//Versions log
//19.02.2020 TOAN, Created

ValueFiltered := k * ValueRaw + (1 - k ) * ValueOld;

ValueOld:= ValueFiltered;
```

The PLC scan time is the sampling time. In practice, **k** must be adjusted, so that signal from the sensor measurements look like the curve which is wanted.

The graph shows filtered signals at different values by the constant k. See table on previous page for explanation of the three signal curves.

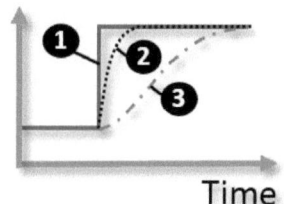

Time

In the next chapter, find a PLC code example.

13.5 Simulation signals for testing of program code

This chapter describes simulation signals, which can be used during development and testing of the programming code. The machine or the hardware panel is often not available, when the PLC code and program needs to be tested. The hardware has possibly not arrived onsite, the machine is not built yet, or the PLC equipment is already shipped to the customer. Therefore, it can be advantageous to be able to simulate 'sensor' signals to verify that the PLC program is working as expected.

Below are programming code examples for four simulation signals, where the frequency and amplitude can be adjusted to meet your requirements.

The signals can be combined to create new simulation signals like this:

```
MySignalWave:= TriangleWave + SineWave;
```

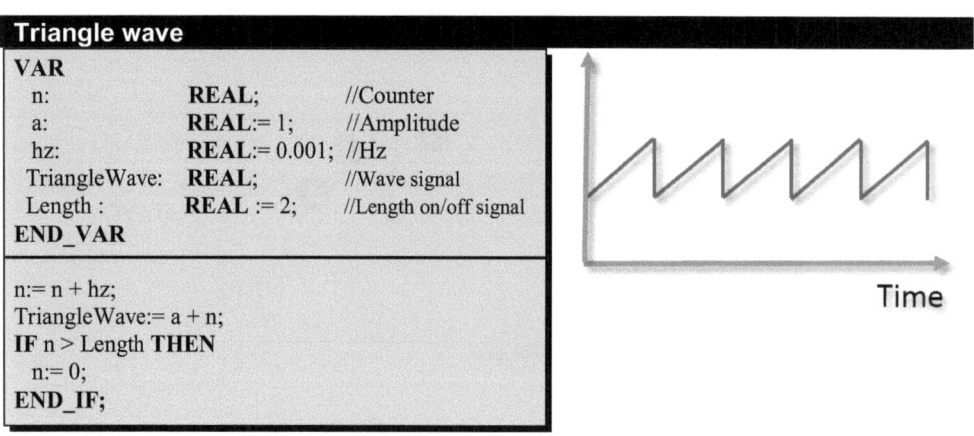

Sine wave

```
VAR
  SineWave:   REAL;              //Wave signal
  n:          REAL;              //Counter
  a:          REAL:= 1;          //Amplitude
  hz:         REAL := 0.001;     //Hz
END_VAR

n:= n + hz;
SineWave:= SIN (n);
```

Triangle wave

```
VAR
  n:           REAL;             //Counter
  a:           REAL:= 1;         //Amplitude
  hz:          REAL:= 0.001;     //Hz
  TriangleWave: REAL;            //Wave signal
  Length :     REAL := 2;        //Length on/off signal
END_VAR

n:= n + hz;
TriangleWave:= a + n;
IF n > Length THEN
  n:= 0;
END_IF;
```

Square wave

```
VAR
  n:            REAL;          //Counter
  a:            REAL:= 1;      //Amplitude
  hz:           REAL:= 0.001;  //Hz
  SquareWave:   REAL;          //Wavesignal
  Length :      REAL := 2;     //Length on/off signal
END_VAR

n:= n + hz;
SquareWave:= 0;

IF n > Length/2 THEN //50% duty cycle
  SquareWave:= a;
END_IF;

IF n > Length THEN
  n:= 0;
END_IF;
```

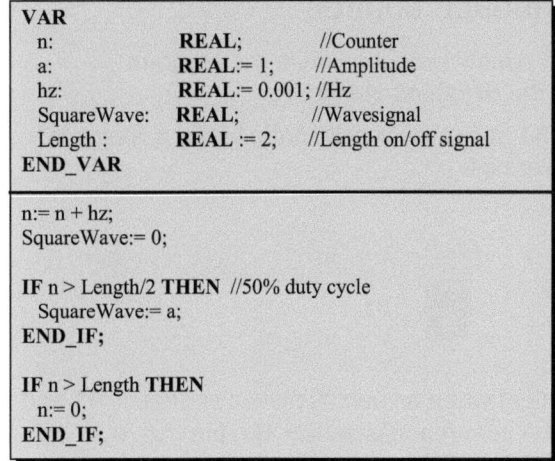
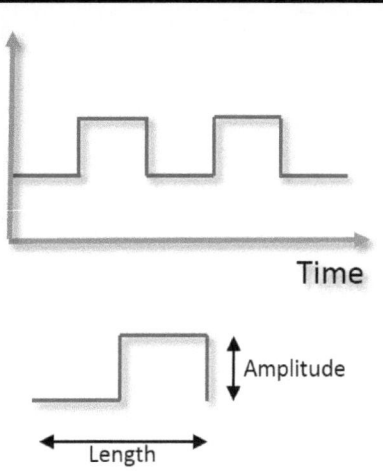

Square wave (filtered)

```
VAR
  n:            REAL;          //Counter
  a:            REAL:= 1;      //Amplitude
  hz:           REAL:= 0.001;  //Hz
  SquareWave:   REAL;          //Wave signal
  Length :      REAL := 2;     //Length on/off signal
  Filter :      LP_Filter;     //Filter Function block *)
  FSquareWave:  REAL;          //Wave signal filtered
END_VAR

n:= n + hz;
SquareWave:= 0;

IF n > Length/2 THEN //50% duty cycle
  SquareWave:= a;
END_IF;

IF n > Length THEN
  n:= 0;
END_IF;

//Filter signal
Filter(ValueRaw:= SquareWave,
       k:= 0.01,
       ValueFiltered => FSquareWave);
```

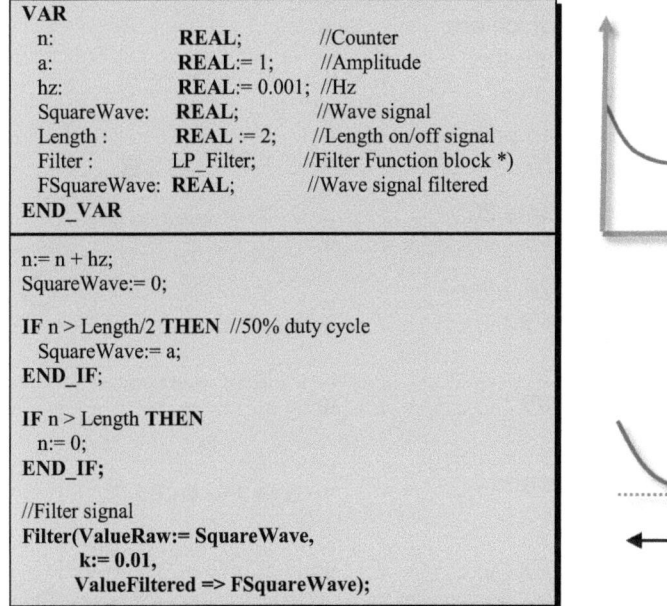
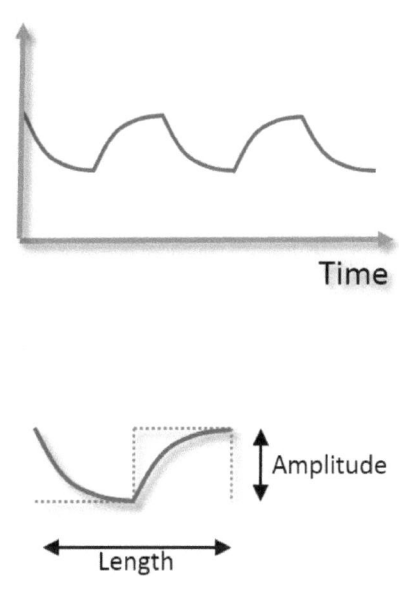

*) Description of the filter function block, see chapter 13.4, page 134.

13.6 Conveyor belt with sequence control

This example describes a conveyor belt control program where the program sequence control is designed following the EN60848 standard.

The example consists of a start switch **S1**, a conveyor belt controlled by a motor **M1** and a sensor at each end of the conveyor belt:

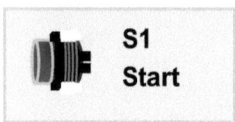
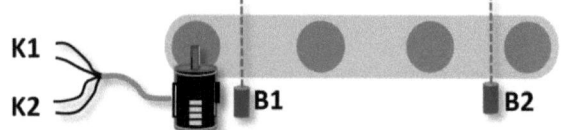

Control Description:
The conveyor belt starts moving to the right when an item is placed at sensor **B1** and the start switch **S1** is activated. When the item reaches sensor **B2**, the conveyor belt stops for 10 seconds. Then the item is pushed back to the left. The conveyor belt stops when the item reaches **B1** again. The motor is controlled by two digital signals: **K1** to start or stop, and **K2** to control the direction of movement (right or left).

The sequence diagram is shown to the right. There are four states in the sequence which are referred to as 0, 10, 20 and 30. These are the four states the program can be divided into.

The start state is 0. When **S1** and **B1** are both **TRUE**, the program changes to state 10.

The state is changed from state 10 to state 20 when the item activates sensor **B2**.

The item waits for 10 seconds in state 20, after which the state is changed to 30, where the item is returned to start.

The entire sequence stops when the item activates sensor **B1** again.

The program is shown on the next page and consists of one main program and two program modules.
In the program a **CASE** statement is used for the sequencer control code, because this gives a good program structure.

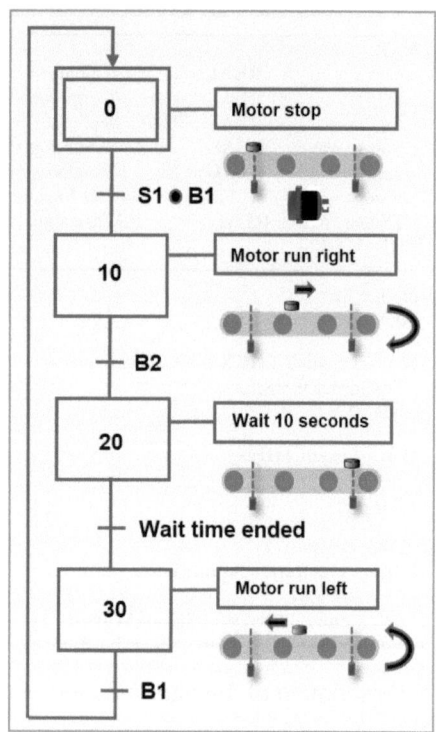

```
PROGRAM ConveyorMain
VAR
  S1 :  BOOL; //Start switch (NO – Normally Open contact)
  B1 :  BOOL; //Belt sensor at start (NO – Normally Open contact)
  B2 :  BOOL; //Belt sensor at end (NO – Normally Open contact)
  K1 :  BOOL := FALSE; //M1, Motor run, 1=run, 0=Stop
  K2 :  BOOL := FALSE; //M1, Motor direction, 1=Right, 0=Left
  Seq : INT := 0; //Sequence no
  B2Delay : TON; //Delay timer at B2 sensor point
END_VAR
```

```
//Program code for ConveyorMain
ConveyorSeq();
ConveyorSetOutPut();
```

```
//ACTION ConveyorSeq – set sequence number
CASE Seq OF
  0 :  IF (S1 AND B1) THEN //Start
         Seq:= 10; //Set to next sequence
       END_IF;
  10 : IF B2 THEN //At the end point
         Seq:= 20; //Set to next sequence
       END_IF;
  20 : B2Delay (IN:= NOT B2Delay.Q, PT:= T#10S); //10 seconds delay
       //Auto reset of the timer ensures it is ready for next start
       IF S2Delay.Q THEN  //Delay ended
         Seq:= 30; //Set to next sequence
       END_IF;
  30 : IF B1 THEN //Item is back at the beginning
         Seq:= 0; //Set to next sequence
       END_IF;
END_CASE;
```

```
//ACTION ConveyorSetOutPut – check the Seq variable and set digital outputs
CASE Seq OF
  0 :  K1:= FALSE;  //Stop motor
  10 : K1:= TRUE;   //Start motor
       K2:= TRUE;   // Set conveyor belt to move right
  20 : K1:= FALSE;  //Stop motor
  30 : K1:= TRUE;   //Start motor
       K2:= FALSE;  //Set conveyor to move left
END_CASE;
```

The variable **Seq** handles the program state and a **CASE** statement is used because this provides a good program structure for a sequencer.

The program module **ConveyorSeq** sets the variable **Seq** to the correct state, based on the current state and the input signals: **S1**, **B1** and **B2**.

The **ConveyourSetOutPut** program module ensures that the two digital outputs (**K1**, **K2**), that controls the motor, are correctly set based on the state of the program.

The program on the page with the sequence control of a conveyor belt only has one start switch **S1**. However, there must be a manual stop switch to stop the conveyor belt if needed. When a conveyor belt stops the program execution, consider the following: Should the item remain on the conveyor belt after it stops, or should the item be returned to the starting point? The choice depends largely on the machine type or plant, because it can be expensive to simply discard items that are on the conveyor belt.

For user operation in this example, these four manual switches are used:

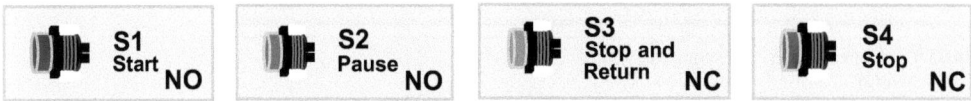

The program is split up into four program sequences which correspond to the switches:

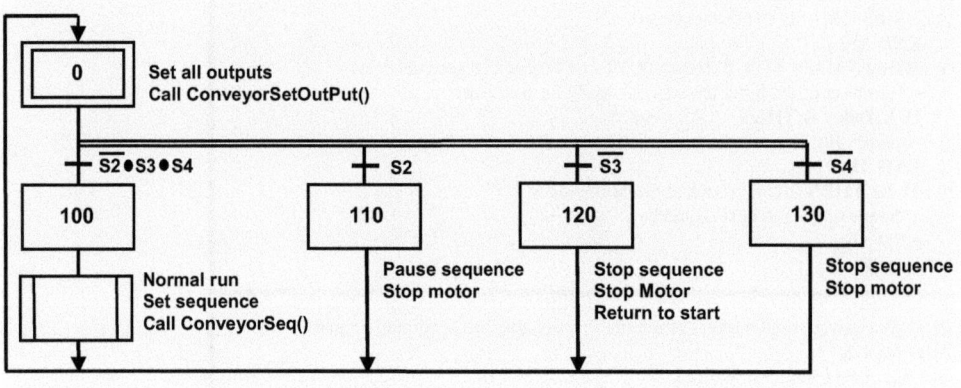

In normal program execution (automatic operation), **S1** must be activated once.

When activating one of the switches **S2**, **S3** or **S4**, each of the states 110, 120 or 130 have their own unique situation which needs to be handled: The conveyor belt is stopped, the sequence is stopped, or the sequence number is reset. In these cases the **ConveyorSeq** program will not be executed.

The program code is found on the next page where the the main sequence is controlled by the **MainSeq** variable. The **SeqSaved** variable is used to save the sequence number when **S2** is activated to pause (put on hold) the **ConveyorSeq** program. Furthermore, the timer in **B2Delay** is changed from the default timer **TON** to **TONP**, because it is not possible to pause a **TON** timer. **TONP** is found on page 125.

```
PROGRAM ConveyorMain
VAR
  S1 : BOOL; //Start switch (NO – Normally Open contact)
  S2 : BOOL; //Pause switch (NO – Normally Open contact)
  S3 : BOOL; //Stop and return switch (NC – Normally Closed contact)
  S4 : BOOL; //Stop switch (NC – Normally Closed contact)
  B1 : BOOL; //Belt sensor at start position (NO – Normally Open contact)
  B2 : BOOL; //Belt sensor at end position (NO – Normally Open contact)
  K1 : BOOL := FALSE; //M1, Motor run, 1=run, 0=Stop
  K2 : BOOL := FALSE; //M1, Motor direction, 1=Right, 0=Left
  Seq, SeqMain, SeqSaved : INT := 0; //Sequence no
  B2Delay : TONP; //Delay timer with pause function
END_VAR
```

```
//Program code for ConveyorMain
MainSeqSelect();
MainSeqEXE();
```

```
//ACTION MainSeqSelect
//Handles the switches and sets the required main sequence number

IF NOT S2 AND S3 AND S4 THEN //Not pause or stop activated. B4 and B3 is NC
  MainSeq:= 100; //Normally run
END_IF;

IF S2 THEN //Pause switch activated
  MainSeq:= 110;
END_IF;

IF NOT  S3 THEN //Stop and return item to start, NC switch
  MainSeq:= 120;
END_IF;

IF NOT S4 THEN //Stop, NC switch (must important and therefore placed last)
  MainSeq:= 130;
END_IF;
```

```
//ACTION MainSeqEXE

ConveyorSetOutPut(); //has to be executed first, because output can be overwritten

CASE MainSeq OF
  100: ConveyorSeq();
       B2Delay(PAUSE:= FALSE);  //Restarts pause. Use TONP function block
  110: SeqSaved:= Seq; //Saves the Seq number
       Seq:= 0; //Set to stop M1 motor
       ConveyorSetOutPut(); //Executes to stop M1 motor
       Seq:=SeqSaved; //Reloads the Seq number
       B2Delay(PAUSE:= TRUE); //Pause. Use TONP function block
  120: Seq := 30; //Sequence that returns to start position
  130: Seq:= 0; //Set to stop M1 motor
END_CASE;
```

13.7 Pump control with two pumps

This example describes a pump control with two pumps and three float switches in a well. The float switches are attached to cables and when the liquid rises in the well, the float switches will be activated and change their NC / NO position.

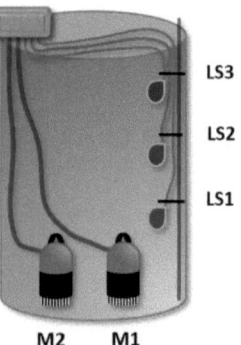

The two submersible pumps **M1** and **M2** have been set to alternate operation mode, which means that they take turns to start and run. With alternating operation, the length of time the pumps are operating is distributed across both pumps, and servicing of the pumps can be performed at the same time.

Signals from the float switces **LS1**, **LS2** and **LS3** determine when the pumps should start or stop. If the level in the well is between **LS1** and **LS2**, one pump must be running.

If the level is above **LS3**, both pumps must be running to pump at full capacity. **LS3** is a **N**ormally **C**losed (NC) switch to provide overflow protection when both pumps are operating, if the **LS3** wire is disconnected or the flow switch is defect. If the level is below **LS1**, both pumps must be stopped to avoid dry run.

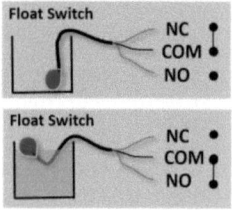

Float switches are connected to digital inputs and the pumps to digital outputs. The following variables are used:

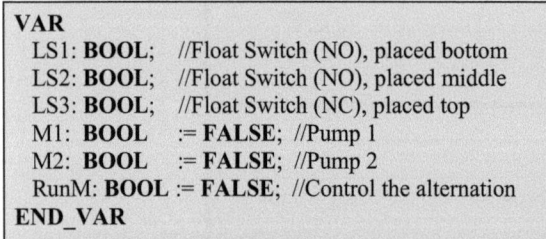

```
VAR
    LS1: BOOL;   //Float Switch (NO), placed bottom
    LS2: BOOL;   //Float Switch (NO), placed middle
    LS3: BOOL;   //Float Switch (NC), placed top
    M1: BOOL    := FALSE; //Pump 1
    M2: BOOL    := FALSE; //Pump 2
    RunM: BOOL := FALSE; //Control the alternation
END_VAR
```

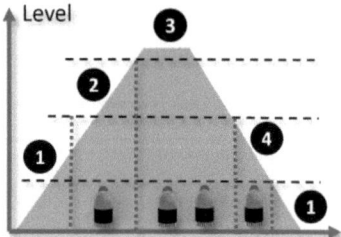

With alternating operation the pumps must take turns to start and run. To select which pump to start, the variable **RunM** is used. If **RunM** is **TRUE**, pump **M1** will start and if **RunM** is **FALSE**, pump **M2** will start. The variable **RunM** selects which of the two pumps that must be turned off when the level is below float switch **LS3**.

The pump variables **M1** and **M2** are by default set to **FALSE** when the PLC powers up, to ensure that the pumps are not running during startup.

The program code is split up into four states defined by the float switches, the number of pumps in operation, and an increased or decreased liquid level:

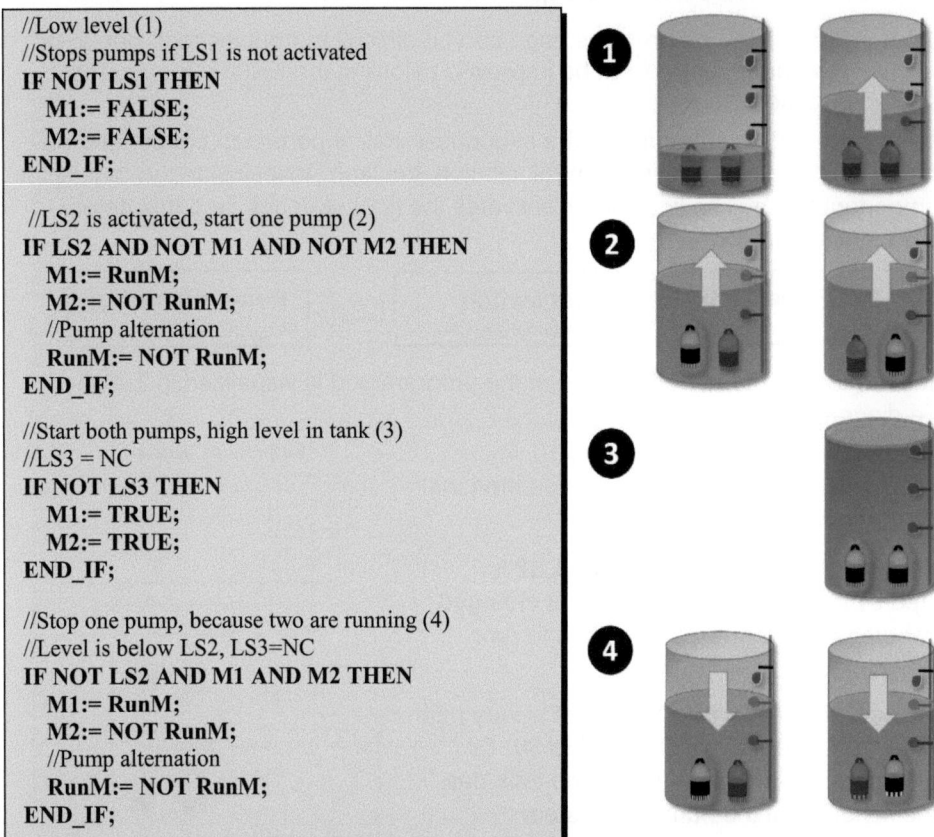

```
//Low level (1)
//Stops pumps if LS1 is not activated
IF NOT LS1 THEN
    M1:= FALSE;
    M2:= FALSE;
END_IF;

    //LS2 is activated, start one pump (2)
    IF LS2 AND NOT M1 AND NOT M2 THEN
    M1:= RunM;
    M2:= NOT RunM;
    //Pump alternation
    RunM:= NOT RunM;
END_IF;

//Start both pumps, high level in tank (3)
//LS3 = NC
IF NOT LS3 THEN
    M1:= TRUE;
    M2:= TRUE;
END_IF;

//Stop one pump, because two are running (4)
//Level is below LS2, LS3=NC
IF NOT LS2 AND M1 AND M2 THEN
    M1:= RunM;
    M2:= NOT RunM;
    //Pump alternation
    RunM:= NOT RunM;
END_IF;
```

The state changes when a float switch is activated or deactivated.

The **IF** statements in state 2 and 4 ensures that the code is executed once only, because the number of running pumps prevents the code to be executed again.

Alarm monitoring for float switch errors:
This code can be implemented to monitor potential float switch errors:

```
//Defect LS2 or LS3 float switch: Because LS3 is activated and LS2 is not activated.
LS2_3_Alarm:= LS2 OR NOT LS3; //LS3 = NC, LS2 = NO

//Defect LS1 or LS2 float switch: Because LS2 is activated and LS1 is not activated.
LS1_2_Alarm:= NOT LS1 OR LS2; //LS1 = NO, LS2 = NO
```

13.8 Pump control with sequence control

The program code from the previous page can be difficult to troubleshoot, debug or extend. The program structure can be improved following the EN60848 sequence control standard (sequencer).

To design the program structure for the sequencer, it is important to understand how the control solution works. In this control solution, the liquid level inside the tank is the most important factor. The liquid level activates the float switches, and this determines the pump operation:

The float switches determine the state of the program and is visualized in the sequence diagram below:

The program changes from state 0 to 10 when **LS1** is activated by the liquid level inside the tank. In state 10 both pumps stop.

When the program is in state 10, and **LS2** is activated by the liquid level, the state is changed to 20 where one of the two pumps has to run.

In state 20, there are two options:
- **LS3** is activated when the liquid level is very high in the tank, and the state is changed to 30. *Or*
- **LS1** is activated when the level in the tank has become so low, that the pump must stop.

The tank is full in state 30 and both pumps must be running. If the liquid level is below LS2, the state must be changed to 20.

To avoid unnecessary start/stop of the pumps caused by waves in the tank, two float switches must be activated before a pump is stopped.

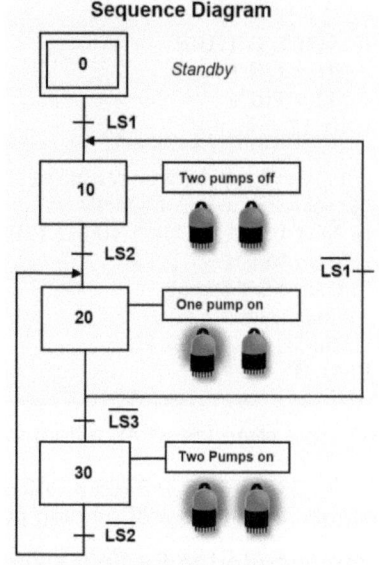

The program is split up into three programs. The **SeqSelect** program checks the signals from the float switches and sets the program state. The **SeqOutput** program starts and stops the pumps depending on the state.

When the program enters state 20, only one of the two pumps should run. The purpose of the **SeqOld** variable is to check, if the pump is entering step 20 for the first time. If it is the first time, the pump which did not run the last time is started.

```
PROGRAM MAIN
VAR
    LS1: BOOL; //Float Switch (NO) bottom, Stop both pumps
    LS2: BOOL; //Float Switch (NO) middle, Run M1 or M2
    LS3: BOOL; //= FALSE; //Float Switch (NC) top, run both pumps
    M1: BOOL:= FALSE; //Pump 1
    M2: BOOL:= FALSE; //Pump 2
    Seq: INT := STEP_0; //Current sequence number
    SeqOld: INT;        //Old sequence number
    RunM: BOOL;         //Controlling the alternation
END_VAR
VAR CONSTANT
    STEP_0: INT := 0;   //Sequence for standby
    STEP_10: INT := 10;
    STEP_20: INT := 20;
    STEP_30: INT := 30; //Sequence for tank is full
END_VAR

//Main Program
SeqSelect(); //Check float Switches and set Sequence
SeqOutput(); //Set the output signals
```

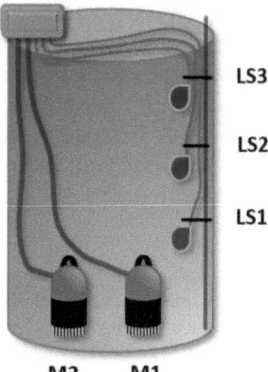

LS3

LS2

LS1

M2   M1

Tank with pumps and
float switches

```
//ACTION SeqSelect

//Save current seq, to handle alternation between pumps
SeqOld:= Seq;

CASE Seq OF
  STEP_0 : //Standby, no liquid in tank
      IF LS1 THEN
        Seq:= STEP_10;
      END_IF;
  STEP_10 :
      IF LS2 THEN   //Level => higher
        Seq:= STEP_20;
      END_IF;
  STEP_20 : //Tank half full
      IF NOT LS1 THEN //Level => lower
        Seq:= STEP_10;
      END_IF;

      IF NOT LS3 THEN //Level => higher
        Seq:= STEP_30;
      END_IF;
  STEP_30 : //Tank is full
      IF NOT LS2 THEN  //Level => lower
        Seq:= STEP_20;
      END_IF;
END_CASE;
```

```
//ACTION SeqOutPut

//Set output variables

CASE Seq OF
  STEP_10 : //Turn pumps off
      M1:= FALSE;
      M2:= FALSE;

  STEP_20 : //Run one pump

      //Alternate between pumps
      //Change pump if first time run
      IF Seq <> SeqOld THEN
        RunM:= NOT RunM;
      END_IF;

      //Turn pumps on or off
      M1:=RunM;
      M2:= NOT RunM;

  STEP_30 : //Tank full
      //Turn both pumps on
      M1:= TRUE;
      M2:= TRUE;
END_CASE;
```

13.9 Automatically and manually operated pump control

The program code from chapter 13.7, page 142, only works for automatically operated pump controls, where flow switches determines whether the pumps should be on or off. However, during servicing and testing the operator needs to be able to turn pumps on and off manually. The example below shows how the program can be split up into automatic and manual operations.

The **Main** program (**PRG**) consists of the following program code and variables:

Main (PRG)	

//Program code for MAIN
IF SwitchAuto AND SwitchOn THEN
  PRGPumpAuto();
END_IF;
 	
IF NOT SwitchAuto AND SwitchOn THEN
  PRGPumpMan();
END_IF;
 	
IF NOT SwitchOn THEN
  PRGPumpOff();
END_IF;
	PROGRAM MAIN
	VAR
	  SwitchM1: **BOOL;** //Manual switch on/off for Motor 1
	  SwitchM2: **BOOL;** //Manual switch on/off for Motor 2
	  SwitchAuto: **BOOL;** //Switch for automatic or manuel
	  SwitchOn: **BOOL;** //Power switch on or off
	  LS1: **BOOL;** //Float Switch (NO) button, stop pumps
	  LS2: **BOOL;** //Float Switch (NO) mid, run M1 or M2
	  LS3: **BOOL;** //Float Switch (NC) top, run both pumps
	  M1: **BOOL**:= **FALSE;** //Pump 1
	  M2: **BOOL**:= **FALSE;** //Pump 2
	  RunM: **BOOL**:= **FALSE;** //Controls the alternation
	END_VAR

The **Main** program (**PRG**) consists of the following program code and variables:

PRGPumpAuto	PRGPumpMan	PRGPumpOff
	//Manual pump on or off **M1:= SwitchM1;** **M2:= SwitchM2;**	//Set pumps to off **M1:= FALSE;** **M2:= FALSE;**

See page 142

The user can set the pump control to automatic or manual operation via a control panel (HMI) as shown to the right.

In automatic operation mode, the **PRGPumpAuto** program will run. If manual operation is selected, the **PRGPumpMan** allows the operator to turn the pumps on or off individually using **SwitchM1** or **SwitchM2**. The position of **SwitchOn** determines whether the **PRGPumpOff** program is running or not.

By splitting the program code into **ACTIONs**, a good program structure is obtained. The example shown is just one of several options for a sensible program split.

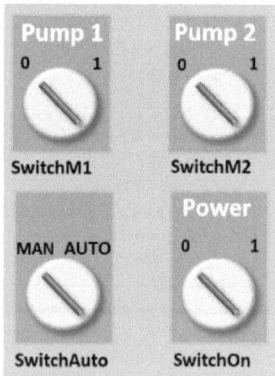

13.10 Calculating tank volume, cylinder on hemisphere

This chapter shows an implementation of a volume calculation for a large storage tank.

The tank consists of a cylinder with a hemisphere in the bottom.

Formulas for calculation of volume are found on the internet.

A function is created where the size tank size are input values, so that the PLC code can be reused for tanks of other sizes. Furthermore, the liquid level is input parameter to the function and the return value is the current volume. The liquid level is measured by an analogue sensor. It can be a pressure sensor in the bottom of the tank, measuring the liquid level or a sensor at the top measuring from the top downwards to the liquid level. The content of the tank is often the main factor to determine which sensor technology should be used. In this solution example, the level is measured from the bottom of the tank up to the liquid surface.

The first aim of this task is to make the tank measurement succeed. Only when the hemisphere is filled, the liquid starts filling the cylinder. The measurements must be checked by a tank calculator found on the internet.

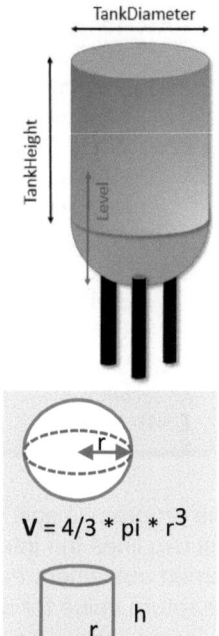

$$V = 4/3 * pi * r^3$$

$$V = pi * r * h^2$$

The calculation is carried out without units to make a more flexible solution, which can be reused. This means that all units must be the same. Units can be measured in meter, ft, cm or mm. The volume becomes cubic: m^3, ft^3, cm^3 or mm^3.

Next, the calculation of the hemisphere must work, and the whole solution must finally be implemented, so that a total test can be made. It is an advantage to document the test in a document to prove that the function has been tested. For the test to be valid, a range of test points must be selected, which must include points outside the range of the tank, at different liquid levels in the tank and close to the interfaces, i.e. where the cylinder and the hemisphere meet. The expected volume is calculated at different levels. It can be made by a calculator or through one of many online pages, where it is possible to make volume calculations on tanks. Finally, the function is tested and the test result is compared with the expected results. (see chapter 16.3, page 194).

Below is shown a suggestion for the PLC code:

```
FUNCTION TankVolumenCal : REAL
VAR_INPUT
    TankDiameter:      REAL;  // Fixed tank diameter
    TankHeight:        REAL;  // Fixed tank height of cylinder
    LevelFromBottom:   REAL;  // Current level measured
END_VAR
VAR CONSTANT
    PI: REAL := 3.1415;
END_VAR
VAR
    Level: REAL;       // Internal calculation
    Vol:   REAL:= 0;   // Internal calculation
    Lr:    REAL;       // Level radius in circle
    TankRadius: REAL;
END_VAR
```

The program is split up in different clear sections as shown on the next page. In the first two lines the internal variables are initialized. Next, the calculation sections are carried out, where each section provides a comment line for information, and finally the return value for the function is set.

A program 'call' for the function could be as follows:

```
Vol:= TankVolumenCal (TankDiameter:= 2,
                      TankHeight:= 6,
                      LevelFromBotton:= LevelSensor);
```

Or like this, because it is a **FUNCTION**:

```
Vol1:= TankVolumenCal (2, 6, LevelSensor);
```

Where **LevelSensor** is the current tank measurement from the bottom.

All values must have the same unit (m, mm, cm, feet, ft).

```
///////////////////////////////////////////////////////////////////////////////////////
//  Tank Volume calculator - Cylinder with a hemisphere
///////////////////////////////////////////////////////////////////////////////////////
Level:= LevelFromBottom;
TankRadius:= TankDiameter/2;

//Check level depth - level cannot be negative
IF Level < 0 THEN
  Level:= 0;
END_IF;

//Check level height - tank cannot be overfilled
IF Level > (TankRadius + TankHeight) THEN
  Level:= TankRadius + TankHeight;
END_IF

//Hemisphere
IF Level <= TankRadius THEN
  // Hemisphere partially filled (1)
  Lr:= SQRT(Level * (TankDiameter - Level));
  Vol:= (PI/6)*level*(3*Lr*Lr+ Level*Level);
ELSE
  // Hemisphere filled (2)
  Vol:= 2.0/3.0*PI * TankRadius * TankRadius * TankRadius;
END_IF;

//Something in the cylinder? (3)
IF Level > TankRadius THEN
   Vol:= Vol + (Level - TankRadius) * PI * TankRadius * TankRadius;
END_IF;

 //Set return value
TankVolumenCal:= Vol;
```

13.11 PLC control for pumping well station with 6 pumps

This example contains an example of a program structure used for a pump well station with 6 pumps. The number of pumps in operation depends on the liquid level inside the well. When the well is full, all pumps must run in order to operate with the highest possible pump capacity.

The pumps are running in alternating operation mode.

Alternating operation mode means:

Pump well with 6 pumps.

1)	The pump with the lowest accumulated operating time must start when higher liquid leves require additional pumping
2)	The pump with the longest accumulated operating time stops when the liquid level becomes lower
3)	After a certain period of time, the pump with the longest accumulated operating time stops, and the pump with lowest accumulated operating time starts.

The level inside the well is measured by a pressure sensor mounted to the bottom of the well. The pressure sensor is an analog sensor signal.

A pump can be switched off by an alarm or manually when service mode. However, these operations are not included in the programming example.

Incorporating alternating operation means that over time the pumps are used for the same amount of time, and ensures that all pumps require maintenance and service at the same time. This reduces overall downtime.

The example can also be used to control e.g. compressor systems, refrigeration plants, power plants or buffer tanks.

When designing the program, principles of Object Oriented Programming (OOP) and the ISA-S88 standard has been incorporated, which means that code for the pumps is collected in a structure which can be reused.

The solution therefore consists of a pumping station using 6 identical pump objects:

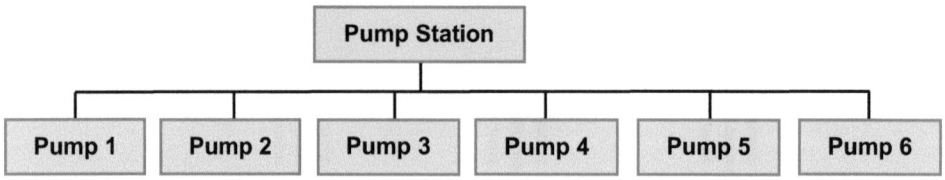

For a pump all variables are collected in a **STRUCT** named **PumpType**:

```
TYPE PumpType :
  STRUCT
  RunState :          PumpState := PumpState.STOP;
  RunTotalMinutes :   DWORD := 0;
  RunLastStartSec:    WORD := 0;
  RunSeconds:         WORD := 0; //Count up to 60
  RunTimerSeconds:    TON; //Count up RunSeconds
  END_STRUCT
END_TYPE
```

In practice, there will be more variables than shown in this example. This could be alarms, power consumption, speed, etc.

Each pump can have different operating modes defined in an **ENUM**:

```
TYPE PumpState :
  ( STOP, RUN, ALARM, SERVICE ) := STOP;
END_TYPE
```

The pumps are declared using an **ARRAY** and used by the pumping station:

```
TYPE PumpStation :
  STRUCT
  PumpsAr:  ARRAY [1..PSCont.MAX_PUMPS] OF PumpType;
  NoPumpsRun:  INT;    //Number of running pumps
  Level:       WORD; //Water level in the well
  END_STRUCT
END_TYPE
```

The pump station also consists of a level sensor and number of pumps in operation.

The **ARRAY** includes a global constant **PSConst** for maximum number of pumps. This ensures that the constant can easily be changed and used throughout the program:

```
TYPE PSConst
VAR_GLOBAL CONSTANT
  MAX_PUMPS : BYTE := 6;
END_VAR
```

The program consists of a main program and 8 functions as shown below:

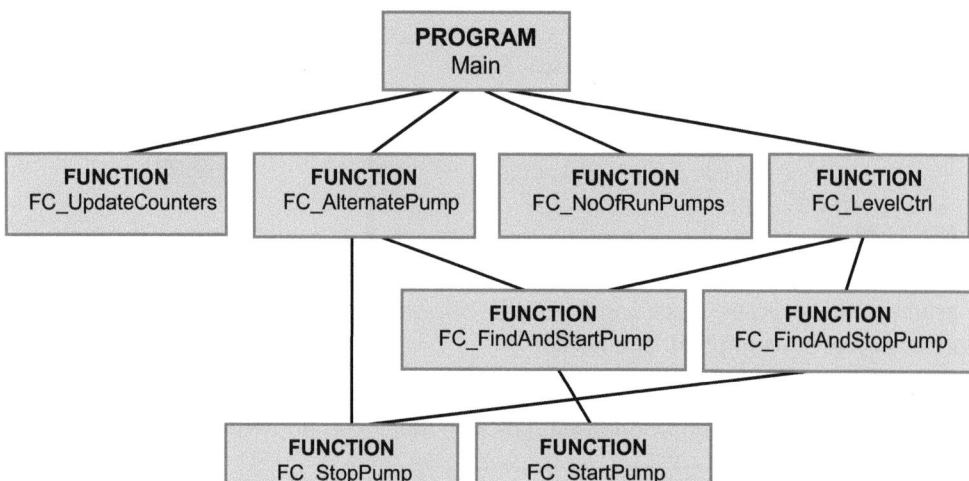

A block diagram provides a good overview of the program structure. If there are many lines and the block diagram is difficult to draw, it may indicate a bad program structure.

The **Main** program uses four functions, and **FC_StopPump** is reused twice. Each function has a maximum of 20-25 program code lines, which is a sign of a good program structure. Functions are used because all variables are stored in a **STRUCT** named **PumpStation**. Using functions make it is easier to change the program structure later on compared to programs split up into **ACTIONS**, because **ACTIONS** always "interconnect" with a program module to exchange variables. Changes to the program structure is often required when adding new features, or when the program needs to be redesigned to redistribute the load on the CPU.

The function **FC_NoOfRunPumps** is not called from **FC_UpdateCounters** because the two functions do not require the same scan time. **FC_UpdateCounters** must be called in each program scan because it uses a **TON** timer, and **FC_NoOfRunPumps** only needs to be called e.g. every second scan time. In the event of later changes to the program, it is therefore better that the functions runs individually.

The following pages cover a proposal for the program code.

```
PROGRAM MAIN
VAR
  PumpWell : PumpStation; //One pumpstation contains all pumps
END_VAR

//Update counters
FC_UpdateCounters(PumpWell.PumpsAr);

//Pump alternating time is 3600 seconds = Hour. Can be change to a lower time when testing
FC_AlternatePump(PumpWell, 3600);

//Update no. of pump runs
PumpWell.NoPumpsRun:= FC_NoOfRunPumps(PumpWell.PumpsAr);

//Find the required pumps to run depending on the level in the pump well
FC_LevelCtrl(751, 1000, 6, PumpWell); //Max level in the pump well station is 1000
FC_LevelCtrl(601, 750, 5, PumpWell);
FC_LevelCtrl(451, 600, 4, PumpWell);
FC_LevelCtrl(301, 450, 3, PumpWell);
FC_LevelCtrl(151, 300, 2, PumpWell);
FC_LevelCtrl(100, 150, 1, PumpWell);
FC_LevelCtrl(0, 99, 0, PumpWell); //To prevent dry runs no pumps must run at a level below level 99
```

Note that the **Main** program has one variable only. This variable is used when calling functions to ensure that the functions can use and change variables.

The maximum liquid level in the pump well station is 1000, and the number of pumps to run at a specified level is determined by the function **FC_LevelCtrl**.

```
FUNCTION FC_NoOfRunPumps : INT
VAR_IN_OUT
  Pumps: ARRAY[1..PSCont.MAX_PUMPS] OF PumpType;
END_VAR
VAR
  i:   INT; //Loop
  No: INT := 0; //Counter
END_VAR

//Count number of running pumps
FOR i:= 1 TO PSCont.MAX_PUMPS DO
  IF Pumps[i].RunState = PumpState.RUN THEN
    No:= No + 1;
  END_IF;
END_FOR;

//Set return value
FC_NoOfRunPumps:= No;
```

```
FUNCTION FC_FindAndStopPump : BOOL
VAR_IN_OUT
  Pumps: ARRAY[1..PSCont.MAX_PUMPS] OF PumpType;
END_VAR
VAR
 i: INT;    RunTime: DWORD;   PumpIndex: INT;
END_VAR
```

```
//Set an initial low time to find the highest run time
RunTime:= 0;
PumpIndex:= 0;

//Find the pump in RUN state with the longest accumulated run time
FOR i:= 1 TO PSCont.MAX_PUMPS DO
  IF RunTime < Pumps[i].RunTotalMinuttes AND Pumps[i].RunState = PumpState.RUN THEN
    RunTime:= Pumps[i].RunTotalMinuttes;
    PumpIndex:= i;
  END_IF;
END_FOR;

//Stop the pump
IF PumpIndex > 0 AND PumpIndex <= PSCont.MAX_PUMPS THEN
  FC_StopPump (Pumps[PumpIndex]);
END_IF;
```

```
FUNCTION FC_FindAndStartPump : BOOL
VAR_IN_OUT
  Pumps: ARRAY[1..PSCont.MAX_PUMPS] OF PumpType;
END_VAR
VAR
 i: INT;    RunTime: DWORD;   PumpIndex: INT;
END_VAR
```

```
RunTime:= 9999999; //Set an initial high time to find a low run time
PumpIndex:= 1;

//Find a pump in STOP state with the lowest run time
FOR i:= 1 TO PSCont.MAX_PUMPS DO
   IF RunTime > Pumps[i].RunTotalMinuttes AND Pumps[i].RunState = PumpState.STOP THEN
     RunTime:= Pumps[i].RunTotalMinuttes;
    PumpIndex:= i;
   END_IF;
END_FOR;

//Start Pump
FC_StartPump(Pumps[PumpIndex]);
```

```
FUNCTION FC_StartPump : BOOL
VAR_IN_OUT
  PumpObj: PumpType;
END_VAR
```

```
//Start Pump
PumpObj.RunState:= PumpState.RUN;
```

```
FUNCTION FC_StopPump : BOOL
VAR_IN_OUT
  Pump: PumpType;
END_VAR
```

```
//Pump will stop after 30 sec delay to avoid unnecessary start/stop
IF Pump.RunLastStartSec > 30 THEN
  Pump.RunState:= PumpState.STOP;
  Pump.RunLastStartSec:= 0;
END_IF;
```

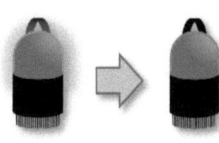

```
FUNCTION FC_UpdateCounters : BOOL
VAR_IN_OUT
  Pumps: ARRAY[1..PSCont.MAX_PUMPS] OF PumpType;
END_VAR
VAR
  i: INT;  //Loop counter
END_VAR
```

```
FOR i:= 1 TO PSCont.MAX_PUMPS DO
  Pumps[i].RunTimerSeconds(IN:= NOT Pumps[i].RunTimerSeconds.Q, PT:= T#1S); //Auto reset

  //Only update when the pump is in RUN state
  IF Pumps[i].RunState = PumpState.RUN THEN

    //Second timer ended?
    IF Pumps[i].RunTimerSeconds.Q THEN
      Pumps[i].RunSeconds := Pumps[i].RunSeconds + 1;
      Pumps[i].RunLastStartSec:= Pumps[i].RunLastStartSec + 1;

      //Update minutes
      IF Pumps[i].RunSeconds >= 60 THEN
        Pumps[i].RunTotalMinutes:= Pumps[i].RunTotalMinutes + 1;
        Pumps[i].RunSeconds:= 0;
      END_IF;

    END_IF; //Second timer end
  END_IF;

END_FOR;
```

```
FUNCTION FC_LevelCtrl : BOOL
VAR_INPUT
 LevelLow: WORD;    //Liquid level in pump well station must be above this level
 LevelHigh: WORD;   // Liquid level in pump well station must be below this level
 NoPumpsReq: INT;   //Number of pumps required to run in the liquid level range
END_VAR
VAR_IN_OUT
 PumpWell : PumpStation; //Address pointer to the STRUCT outside the function
END_VAR
VAR
 PumpCtrl: INT := 0;
END_VAR
```

```
//This function starts or stops a pump depending on the level in the pump well station

//Check level. Does a pump need to be started due to a high liquid level?
IF (PumpWell.Level >= LevelLow) AND (PumpWell.Level <= LevelHigh) THEN
  IF NoPumpsReq > PumpWell.NoPumpsRun THEN
    //Negative if one more pump is needed
    PumpCtrl:= NoPumpsReq - PumpWell.NoPumpsRun ;
  END_IF;
END_IF;

//Check level. Does a pump need to be stopped due to a low liquid level?
IF (FC_LevelCtrl = 0) AND (PumpCtrl = 0) THEN
   IF (PumpWell.Level >= LevelLow) AND (PumpWell.Level <= LevelHigh) THEN
    IF NoPumpsReq < Pit.NoPumpsRun THEN
      //Positive if a pump has to be stopped
      PumpCtrl:= NoPumpsReq - PumpWell.NoPumpsRun;
    END_IF;
   END_IF;
END_IF;

//If a pump needs to be stopped, stop the pump
IF PumpCtrl < 0 THEN
  FC_FindAndStopPump(Pumps:= Pit.PumpsAr);
END_IF;

//If a pump needs to be started, start the pump
IF PumpCtrl > 0 THEN
  FC_FindAndStartPump(Pumps:= PumpWell.PumpsAr);

END_IF;
```

```
FUNCTION FC_AlternatePump : BOOL
VAR_IN_OUT
  PumpWell: PumpStation; //Address pointer to the STRUCT outside the function
END_VAR
VAR_INPUT
  AlternationTimeSec: DWORD; //The time a pump has to run before alternating
END_VAR
VAR
  i: INT; //Counter for LOOP
END_VAR
```

```
//Function for alternating between pumps. Time until pump alternation is in AlternationTimeSec

//Only alternate between pumps if at least one, but not all pumps are running
IF PumpWell.NoPumpsRun > 0 THEN
  IF PumpWell.NoPumpsRun < PSCont.MAX_PUMPS THEN
    //Check all pumps
    FOR i:= 1 TO PSCont.MAX_PUMPS DO
      IF (AlternationTimeSec < PumpWell.PumpsAr[i].RunLastStartSec ) THEN
        IF PumpWell.PumpsAr[i].RunState = PumpState.RUN THEN

        //First stop the pump that has been running for the longest time
        FC_StopPump(PumpWell.PumpsAr[i]);
        //Then start another pump
        FC_FindAndStartPump(PumpWell.PumpsAr);

        END_IF;
      END_IF;
      EXIT; //Only alternate to one pump. Another pump will be checked in next scan
    END_FOR;
  END_IF;
END_IF;
```

 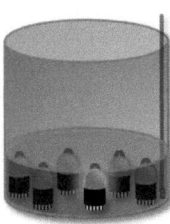   

13.12 EXAMPLE: Heating of liquid in a tank

This example shows a tank control, which can heat liquid.

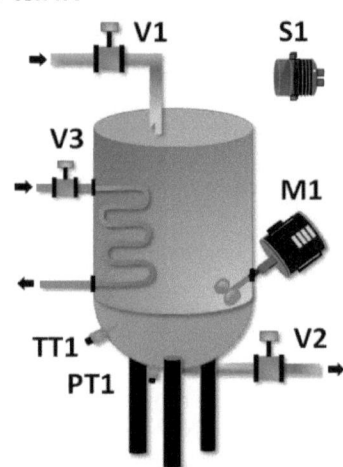

Control Description:

When switch **S1** is activated, valve **V1** opens and liquid fills the tank. When the liquid reaches the requested level, valve **V1** closes. Then valve **V3** opens and heats up the liquid. When the liquid inside the tank reaches the requested temperature, valve **V2** opens to empty the tank.

A mixer **M1** ensures stirring during the heating process.

Description of the used components and their mode of operation:

Name	I/O	Component	Mode of operation:
S1	DI	Start switch	Activating the start switch will start the tank control program sequence. Only starts when the tank is empty.
TT1	AI	Sensor	Temperature Transmitter. Measures temperature in the tank. Measuring range is 0 to 100 degrees. 4-20 mA signal. Connected to an analog input module. Range is 0 to 65335
PT1	AI	Sensor	Pressure transmitter. Measuring pressure. Measures liquid level inside the tank Measuring range is 0 to 2 meters. 4-20 mA signal Connected to analog input module. Range is 0 to 65335
V1	DO	Valve	Signal from PLC opens the valve, and the tank fills up with liquid. No signal closes the valve automatically
V2	DO	Valve	Signal from PLC opens valve and the liquid leaves the tank.
V3	DO	Valve	Signal from PLC opens the valve and hot water is circulated in a spiral tube inside the tank. This causes the liquid inside the tank to heat up. The mixer has to run during this operation
M1	DO	Mixer / Stirrer	Powered by a motor with a frequency converter. The mixer speed is set directly on the frequency converter. Do not turn on the mixer when the tank is empty, as this will cause the mixer to overheat. Controlled by a digital signal from the PLC

The three valves **V1**, **V2** and **V3** close automatically when they no signals (24V).

Design and structure of the PLC program

To ensure a good program structure, the program is designed according to the EN60848 standard, which describes how a program can be designed sequentially. This means that the program is controlled by states (sequences).

Referring to the control description (previous page), this control has four states:

- Wait until the start switch **S1** is activated
- Fill up the tank with liquid
- Heat the liquid inside the tank
- Empty the tank

The program design and structure is documented with a sequence diagram following the EN60848 standard:

The first state is 0, where the control waits for the start switch **S1** to be activated.

When **S1** is activated, the state is set to 10, where valve **V1** is open and the tank will be filled up with liquid.

When the tank has the requested amount of liquid, the state is set to 20. The requested amount of liquid is obtained when the measured tank volume is greater than the user defined value set by the **SetVolume** variable.

Step 20 is the heating process. **V3** must be open and the mixer turned on.

A temperature sensor continuously measures the temperature inside the tank. The progress to step 30 occurs when the temperature is higher than the temperature set by the user.

In step 30 the tank is emptied. When the tank is empty, change to step 0 to restart the control system.

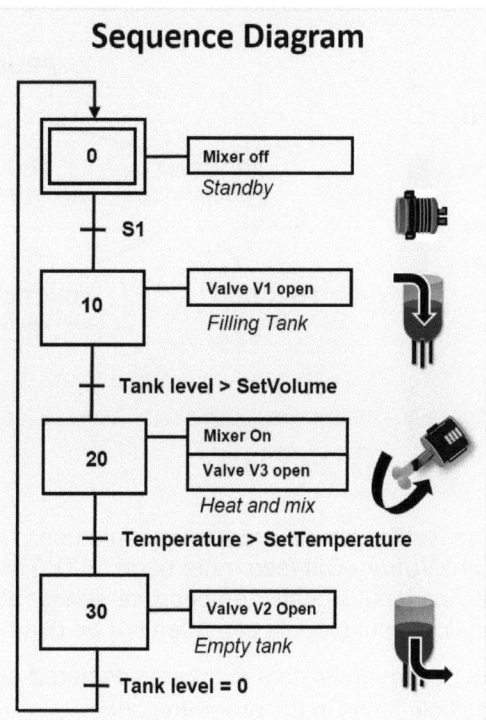

Sequence Diagram

- 0 — Mixer off / Standby
- S1
- 10 — Valve V1 open / Filling Tank
- Tank level > SetVolume
- 20 — Mixer On / Valve V3 open / Heat and mix
- Temperature > SetTemperature
- 30 — Valve V2 Open / Empty tank
- Tank level = 0

Next follows the program design, where the control is split up into seperate functions and program modules.

The program structure has one program module which handles the input signals, and one program module which handles the output signals. This ensures a good and clear program structure. Program modules have been chosen instead of functions, because the code inside the program modules is unlikely to be reused and program tests are more difficult to carry out, if variable sensors are created inside a function.

In addition, a program module ensures that values are written correctly to the HMI. In the program example below, the remaining values for the HMI comes from the **Main** program.

The program is split up as shown in the block diagram below:

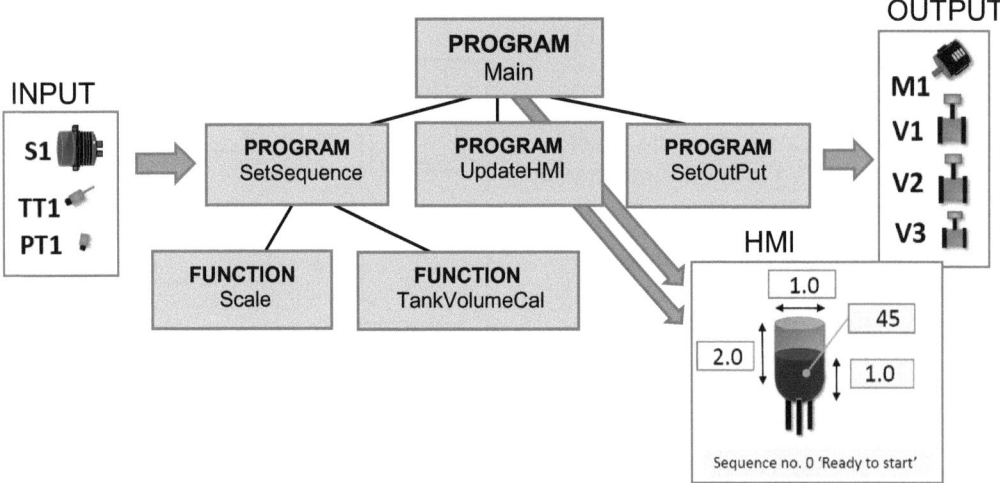

The **SetSequence** program module uses two functions: **Scale** (see more 92) and **TankVolumeCal** (see more page 147). The **Scale** function is used to scale the analog input signals (temperature sensor and level sensor), as the value from an analog input module can often not be used directly.

Each sequence state number is declared as a **CONSTANT** because they are used multiple times in the program code.

The user control panel (HMI) shows the sequence state number, so a service technician can easily see the current sequence state number of the program.

The following pages contains a proposal for the program code.

© 2020 Tom Mejer Antonsen

```
PROGRAM MAIN
VAR
  Seq:              WORD := STEP_0_START;
END_VAR
VAR
  SetTankDiameter: WORD := 1;    //User settings
  SetTemperature:  WORD := 45;   //User settings
  SetVolume:       WORD := 1;    //User settings
  SetTankHeight:   WORD := 2;    //User settings
END_VAR
VAR CONSTANT
  STEP_30_EMPTY : WORD := 30;  //Sequence number
  STEP_20_HEAT  : WORD := 20;  //Sequence number
  STEP_10_FILL  : WORD := 10;  //Sequence number
  STEP_0_START  : WORD := 0;   //Sequence number
END_VAR
```

```
//Main Program
SetSequence(TemperatureRequest:= SetTemperature,
            VolumeRequest:= SetVolume, //M3 volume
            TankDiameter:= SetTankDiameter,
            TankHeight:= SetTankHeight,
            Seq:=Seq);
UpdateHMI (Seq:= Seq);
SetOutPut (Seq:= Seq);
```

```
PROGRAM UpdateHMI
VAR_INPUT
  Seq:           WORD;
END_VAR
VAR
  SegStr:        STRING;
  HMISeqStr: STRING;  //String to HMI
END_VAR
```

```
//Update information to HMI
//Create the string to HMI
HMISeqStr:= CONCAT('$22 Sequens no. $22 ', WORD_TO_STRING (Seq)); // Where $22 is "
HMISeqStr:= CONCAT(HMISeqStr, ' ');

//Show right text depend of the step
CASE Seq OF
  MAIN.STEP_0_START   : HMISeqStr:= CONCAT(HMISeqStr, 'Ready to Start');
  MAIN.STEP_10_FILL   : HMISeqStr:= CONCAT(HMISeqStr, 'Filling Tank');
  MAIN.STEP_20_HEAT   : HMISeqStr:= CONCAT(HMISeqStr, 'Heat and mixing Tank');
  MAIN.STEP_30_EMPTY : HMISeqStr:= CONCAT(HMISeqStr, 'Empty Tank');
END_CASE
```

```
PROGRAM SetSequence
VAR
  S1 :  BOOL;        //Manual start button. Value from INPUT module
  TT1 : WORD;        //Temperature sensor. Value from INPUT module
  PT1 : WORD;        //Pressure sensor. Value from INPUT module
  Vol:  REAL;        //Internal calculation
END_VAR
VAR_INPUT
  TemperatureRequest: REAL;
  VolumeRequest:       REAL;
  TankDiameter:        REAL;
  TankHeight:          REAL;
END_VAR
VAR_IN_OUT
  Seq:                 WORD;
END_VAR
```

```
//Always calculate to avoid long lines of code
Vol:= TankVolumenCal(TankDiameter,TankHeight, Scale(PT1, 0, 65535, 0, 2));

CASE Seq OF
  MAIN.STEP_0_START :  IF S1 THEN  //Only allow to use S1 in step 0
                         Seq:= MAIN.STEP_10_FILL;
                       END_IF;
  MAIN.STEP_10_FILL  : IF Vol > VolumeRequest THEN
                         Seq:= MAIN.STEP_20_HEAT;
                       END_IF;
  MAIN.STEP_20_HEAT :  IF Scale(TT1, 0, 65535, 0, 100) > TemperatureRequest THEN
                         Seq := MAIN.STEP_30_EMPTY;
                       END_IF;
  MAIN.STEP_30_EMPTY : IF (Vol < 0.001) THEN  //In case the  REAL calculation is not zero
                         Seq:= MAIN.STEP_0_START;
                       END_IF;
END_CASE;
```

```
PROGRAM SetOutPut              //PROGRAM SetOutPut
VAR_INPUT                      //Reset all output. CASE Sets the correct output value
  Seq: WORD;                   M1:= FALSE; V1:= FALSE; V2:= FALSE; V3:= FALSE;
END_VAR
VAR                            CASE Seq OF
  M1: BOOL; //Mixer              MAIN.STEP_10_FILL    : V1:= TRUE;  //Fill
  V1 : BOOL; //Valve to fill     MAIN.STEP_20_HEAT  : M1:= TRUE;  //Mixer
  V2 : BOOL; //Valve to empty                         V3:= TRUE;  //Heat
  V3 : BOOL; //Valve to heat     MAIN.STEP_30_EMPTY : V2:= TRUE;  //Empty
END_VAR                        END_CASE;
```

13.13 EXAMPLE: FC Toggle switch (two-way switch)

This example shows a function block that can be used as a Toggle Switch.
A Toggle Switch changes status each time the switch is activated, and can be used as a start and stop switch for an electrical component (eg motor, fan or light).

The advantage of a Toggle Switch is that you only need to use one switch, instead of using a switch for on and a switch for off.

To the right, a time diagram is shown. The signal from the switch is **CLK** and the component to be switched off or on must be connected to **Q**. **CLK_OLD** is an internal variable used as a OneShot and ensures that there is only one change in **Q** status, every time **CLK** is activated.

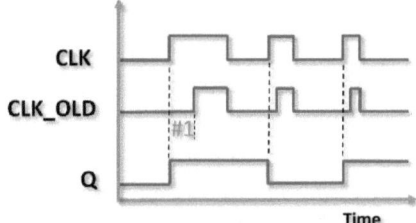

The example is based on the do-it-yourself implementation of the Oneshot, (see page 111), which ensures that the solution can be used in all types of PLCs:

```
FUNCTION_BLOCK FBToggle
VAR_INPUT
  CLK: BOOL; // Input signal
END_VAR
VAR
  // Remember previous signal on CLK
  CLK_OLD: BOOL := FALSE;
END_VAR
VAR_OUTPUT
  Q: BOOL; // Output
END_VAR
```

```
//Code for FUNCTION BLOCK: Toogle
//When CLK is moves from FALSE to TRUE
//Q will be TRUE if FALSE or FALSE if TRUE

//Detect the rising edge on the input signal
IF CLK AND NOT CLK_OLD THEN
    CLK_OLD:= TRUE;  //#1
    Q:= NOT Q;  //#2
END_IF;

//Reset the rising edge detection
IF NOT CLK THEN
    CLK_OLD:= FALSE;
END_IF;
```

Change of status is made at **#2**, where **Q** is set to the inverted value of **Q**.

Program example:

```
PROGRAM MAIN
VAR
  MyToogle: FBToggle;
  K1:        BOOL := FALSE;
  S1:        BOOL; //Contact switch
END_VAR

//Example program:
MyToogle (CLK:= S1, Q=> K1);
```

13.14 EXAMPLE: 3D car park controlled by a robot

This chapter shows an example of a PLC program used to handle cars inside a car park.

When the car is inserted into the car park house by the robot, the program has to first find a vacant space to ensure the robot knows where to place the car.

When the car is picked up by the car driver, the car must first be located by the program, so the robot knows where to pick up the car from.

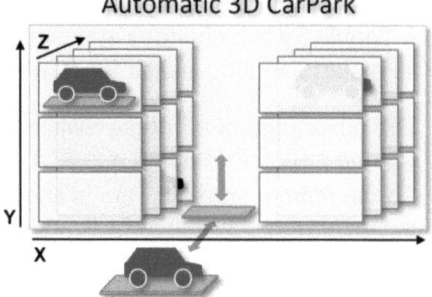

Automatic 3D CarPark

Due to the fact that a car license plate is unique, it is used as the identifier, and as it contains numbers and letters it is set to a **STRING** data type in the PLC program. The length of the **STRING** is limited to 15 characters to save memory in the program. The constant for this is named **STR15**.

It is possible to insert, find or remove cars from the parking house. To reuse as much of the program code as possible, only one function named **CarHandle** is implemented which takes one of the following parameters: CAR_INSERT, CAR_FIND or CAR_DEL.

A location inside the 3D parking house consists of x, y and z coordinates and therefore a **STRUCT** named **Pos** is declared because this gives a clear structure to the code.

```
TYPE Pos :
  STRUCT
    x: INT;
    y: INT;
    z: INT;
  END_STRUCT
END_TYPE
```

```
VAR_GLOBAL CONSTANT
  STR15 : INT:= 15; //Car number
  CAR_INSERT : INT := 1;
  CAR_FIND :    INT:= 2;
  CAR_DEL :     INT:= 3;
END_VAR
```

```
VAR_GLOBAL CONSTANT
  X_MAX: WORD := 2;
  Y_MAX: WORD := 3;
  Z_MAX: WORD := 4;
END_VAR
```

Below see how the **CarHandle** function can be used:

```
PROGRAM Main
VAR
  MyPos: Pos; //Location inside the car park house
  ArCarPark: ARRAY[1.. GVL.X_MAX, 1.. GVL.Y_MAX, 1.. GVL.Z_MAX] OF STRING[GVL.STR15];
END_VAR
```

```
// REMEMBER TO run the code only once!
ArCarPark [1,3,1]:= 'YD 12345'; //Insert car directly into the 3D Array
CarHandle ('YD 12345', GVL.CAR_FIND, ArCarPark, MyPos); //MyPos.x = 1, MyPos.y =3, MyPos.z = 1
CarHandle ('AB 12345', GVL.CAR_INSERT, ArCarPark, MyPos);
CarHandle ('AB 12345', GVL.CAR_DEL, ArCarPark, MyPos);
```

```
FUNCTION CarHandle : BOOL
VAR_INPUT
  CarStr:  STRING; //Number plate for the car
  Handle:  INT;      // What action to take? CAR_FIND, CAR_DEL or CAR_INSERT
END_VAR
VAR_IN_OUT
  arPark:   ARRAY[*, *, *] OF STRING [GVL.STR15]; //Pointer to 3D ARRAY
  CarP:     Pos; // Location of the car
END_VAR
VAR
  Loop, Found : Pos; //Working STRUCT
  Ctrl:  BOOL := FALSE; //Control the operation inside this function. If FALSE an error occur
END_VAR
```

```
//Copy ARRAY sizes to local variables to maintain readable program code for the LOOPs
Loop.x:= DINT_TO_INT (UPPER_BOUND (arPark, 1)); //Size 1D
Loop.y:= DINT_TO_INT (UPPER_BOUND (arPark, 2)); //Size 2D
Loop.z:= DINT_TO_INT (UPPER_BOUND (arPark, 3)); //Size 3D

//LOOP through all locations in the 3D car park house
FOR CarP.x:= 1 TO Loop.x DO
  FOR CarP.y := 1 TO Loop.y DO
    FOR CarP.z:= 1 TO Loop.z DO

    //Condition only passes when no action has been taken
    IF  Ctrl = FALSE THEN
      IF FIND (ArPark [CarP.x, CarP.y, CarP.z], CarStr) > 0 THEN
        CASE Handle OF
          GVL.CAR_FIND  : Found:= CarP; //Copy coordinates where the car was located
          GVL.CAR_DEL   : ArPark [CarP.x, CarP.y, CarP.z]:= ''; //Set STRING to empty
        END_CASE
        Ctrl:= TRUE; //done
      END_IF;

      //Insert the car at the first free location in the parking house
      IF Handle = GVL.CAR_INSERT AND ArPark [CarP.x, CarP.y, CarP.z] = '' THEN
        ArPark [CarP.x, CarP.y, CarP.z]:= CarStr; //Insert
        Ctrl:= TRUE; //done
      END_IF;
    END_IF;

    END_FOR; //z
  END_FOR; //y
END_FOR;  //x

//Set return values (copy x, y, z coordinates inside STRUCT) and error code
CarP:= Found;
CarHandle := Ctrl;
```

13.15 EXAMPLE: Configurable car wash control

This example describes a program design for a car wash, where it is possible to configure the washing programs via a user panel (HMI) as shown below:

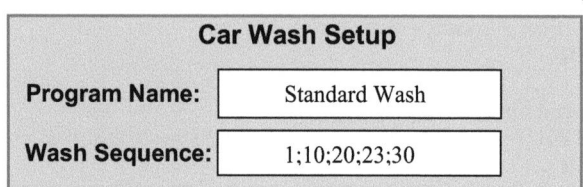

The aim of this program structure is to enable the owner of the car wash to change the washing program him/herself according to the season, without using a PLC programmer to change the program code. It may be necessary to change the washing programs due to the seasonal changes between the summer and winter period. The owner can also change, name or add washing program him/herself.

Each car washing program consists of a number of washing sequences (a recipe), and each washing sequence is configured by numbers and a semicolon. The numbers in the washing sequence refer directly to the programs the PLC executes to run a complete program. The washing sequences are executed in the order it is configured, first 1, then 10 and then 20, etc. Here is a list of selected car washing sequences which can be used to configure a complete car wash program:

Number	Program name	Mode of operating
0	-	Standby. Idle mode. No program to execute
1	WaterOn();	Water open. Move swing arm forward and backward
10	SoapStandard();	Standard soap. Move swing arm forward and backward
11	SoapBudget();	Budget soap. Move swing arm forward and backward
20	Brushing();	Using brushes. Move swing arm forward and backward
22	UnderCarWash();	Washing under the car. Only use water
23	WheelWash();	Wheel brushing. Use soap. Use water
24	NumberPlateWash();	Number plate washing. Front and back
30	Drying();	Drying. Moving swing arm forward and backward.
99	Stop();	Close water tap. Move swing arm to start position. Stop program.

More washing sequences can easily be added, and it is possible to execute the same washing sequence more than once, or reuse it in other washing programs.

When designing the program, there are several ways to ensure a good program structure. Various solutions are discussed below:

Function (FC)

A function cannot be used solely, because the washing program must use a timer (TON) to handle how long a washing sequence needs be carried out.
A timer can de declared and used as a global shared timer, but this is not good program structure as it does not create the function as an independent program, which is the purpose of a function.

Function Block (FB)

Could be used. However, there are many IO signals in a car wash and therefore many variables to handle in each function block. A **STRUCT** could be used to handle the many variables, however, as the overall program of the car wash is a small program, using **FUNCTION BLOCK**s can too complex.
A function block can handle the timers needed for washing timers.

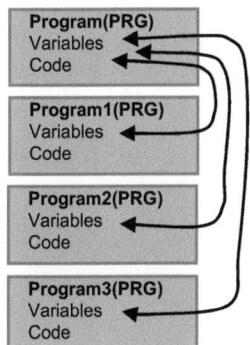

Program(PRG)

Could be used. But all IO variables must be made global, which does not provide a good program structure.

The **Program (PRG)** can share variables with other **Program (PRG)** directly, but it does not provide a good program structure, because all variables are then "horizontally" shared between the program modules as shown on the diagram to the left.

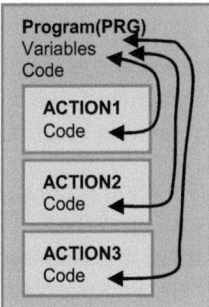

ACTION

The best solution. An **ACTION** only contains pro-gram code and no variables.

All shared variables and IO signals from sensors, valves and motors are declared in the Program (PRG) which allows the variables to be used by the **ACTION**.

The Program (PRG) can include a shared timer for managing wash times.

To ensure a good program structure, all variables for a wash sequence are collected in a **STRUCT**, as shown below:

```
TYPE CarWashType:
   STRUCT
   ProgramName :      STRING;      //Can be a name like 'Standard Wash'
   ProgramNumbers : STRING;       //A sequence such as '1;10;22;24'
   Cost :             REAL := 0;   //Cost off a wash
   NoRuns :           DWORD := 0; //Number of washes
   END_STRUCT
END_TYPE
```

When all the variables are collected in a **STRUCT**, only one **ARRAY** needs to be declared, which can contain all the desired washing programs.

The declared **ARRAY** must be marked with **RETAIN** so all the washing programs, the user has created are saved when restarting the PLC.

To make the program test fast and easy, three washing programs are created:

```
//ACTION ConfigWashPrograms
//Config demo wash programs
ArCarWash [1].ProgramName := 'Budget Wash';
ArCarWash [1].ProgramNumbers := '1;11;20;';
ArCarWash [1].Cost:= 10;

ArCarWash [2].ProgramName := 'Standard Wash';
ArCarWash [2].ProgramNumbers := '1;10;20;30';
ArCarWash [2].Cost:= 15;

ArCarWash [3].ProgramName := 'Gold Wash';
ArCarWash [3].ProgramNumbers := '1;20;10;20;21;24;23;30';
ArCarWash [3].Cost:= 30;
```

Note: This program code can be optimized by using a function. See more in chaper 11.5, page 106.

The program code can be found on the next page, where demo washing program 3 is used. This is the '**Gold Wash**'.

Activating **S1_Start** starts the program. After starting the program, the first number in the washing program sequence is found by using the **GetProgramNo** function (se see more in chapter 11.4, page 103). The first program in the sequence is **WaterOn()**;

The **CASE** statement uses the number found to execute the washing program. Before a washing program is completed, **CarWashRun** must count one up. When **CarWashRun** has added one to the counter, the **GetProgramNo** is called again to find the next washing program number. The loop stops when **GetProgramNo** returns 0 which indicates no number found, or when the user activates **S2_Stop**.

```
PROGRAM MAIN
  VAR RETAIN
  ArCarWash:          ARRAY[1..5] OF CarWashType; //The configuration of washing programs
  END_VAR
  VAR
  S1_StartOneShot:    R_TRIG; //Oneshot for start button
  CarWashSeq:         INT := 0; //Current program found in the STRING: '1,10,20…'
  S1_Start:           BOOL;    //Start button
  CarWashPrg:         INT;     //The user selected washing program to run
  CarWashRunOld:      INT := 0; //The previous index number of ProgramNumber
  CarWashRun:         INT;     //Index number of ProgramNumber. Start with 1, then 2, then 3…
  S2_Stop:            BOOL;    //A stop button
  PrgTimer:           TON;     //A timer shared between all programs. Can be used as wash timer
  END_VAR
```

```
S1_StartOneShot(IN:= S1_Start); //Press for start

IF S1_StartOneShot.Q AND CarWashRun = 0 THEN //Start new washing program
  CarWashPrg:= 3; //User has selected washing program number 3
  CarWashRun:= 1; //Run the first program wash sequence
  ConfigWashPrograms();  //Predefined washing programs for demo and test purposes
END_IF;

IF CarWashRun <> CarWashRunOld THEN //Get new wash program sequence number
  CarWashSeq:=GetProgramNo (CarWashRun,  ArCarWash[CarWashPrg].ProgramNumbers);
  CarWashRunOld:= CarWashRun; //Update previous so any change can be noted
END_IF;

IF S2_Stop THEN CarWashSeq:= 99; END_IF; // Stop button activated

CASE CarWashSeq OF
  //Add one to CarWashRun when a program ends
  0  : CarWashRun:= 0; CarWashRunOld:= 0; //No more sequences to run
  1  : WaterOn();
  10 : SoapStandard();
  11 : SoapBudget();
  20 : Brushing();
  21 : WheelWash ();
  22 : UnderCarWash();
  23 : WheelWash();
  24 : NumberPlateWash();
  30 : Drying();
  99 : Stop();
ELSE
  CarWashSeq:= CarWashSeq *-1; //Error program number not found
END_CASE;
```

13.16 EXAMPLE: Adapt pump speed to save energy

This example describes the design of a PLC program where the speed of a pump is adjusted to keep it running at the most optimal speed. A pump will usually be most efficient when running at 100% speed (maximum capacity), but wear and tear on the pump hose, and changes in fluid form (velocity) will influence the pumps ability to stay at the most optimal speed. Therefore, to save energy the PLC program needs to check whether the speed of the pump is at the most optimal and change it if this is not the case.

The example consists of the following components:

    - Flow meter (1) measuring the amount of liquid running through a tube.
    - Pump (3) with a motor and frequency converter (4).

The components connected to the PLC are shown below:

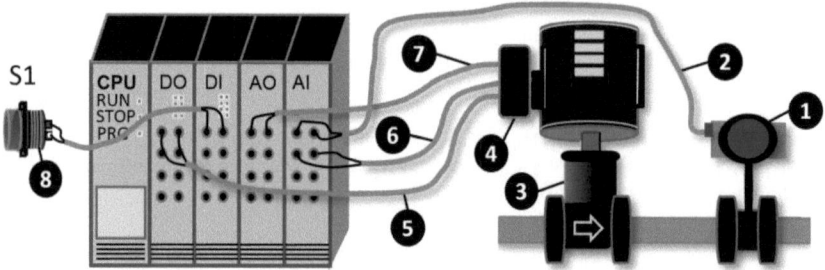

Component overview and mode of operation:

No	Signal	Component
1		Flow Meter. Liquid velocity measurement in [m3/h]
2	4-20 mA	Signal cable from flow meter to AI (Analog Input) on the PLC
3		Pump run by a motor
4		Frequency converter for controlling the speed of the motor
5	Digital	DO (Digital Output) from PC to start and stop the motor
6	4-20 mA	Power measurement [kW] of the motor energy consumption. Connected to AI (Analog Input) on the PLC
7	4-20 mA	Motor Speed Control. Percent signal, 0 to 100%
8	Digital	Manual switch S1. Starts the measuring period

In order to find the most optimal speed for the pump, a number of flow and power measurements is carried out at different pump speeds.

For each measurement:
(A) Set a frequency converter speed and start the pump if the pump has not already started. It is possible to change the speed directly without stopping and starting the pump.

(B) After a period of time, the flow measurement (curve C) is stable. At point (B) measure the flow (curve C) and the power [w] (curve D).

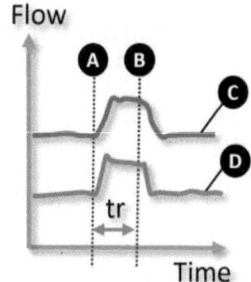

The time between point (A) and point (B) is in seconds, and is found by testing the solution in practice, because the measurements is affected by the pump size, the tube diameter and the reaction time of the flow.

The program code example does not detect when the flow is stable. In this program code example, a fixed time at point (B) is used. To improve the measured result, averages of measurement points could be calculated or a digital filter used.

The function block is named **DataCollect**. The function block uses a timer to set the time interval from (A) to (B). It is important to call the function block in each program scan, because the function block uses a timer.

The flow and power signals are input values to the function block. When the output variable **Done** is **TRUE**, data is ready in the **ValueResult** variable. Thereafter, the frequency converter can be set to a new speed and a new measurement period starts.

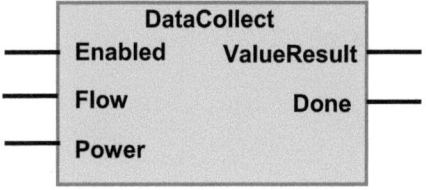

The calculation of **ValueResult** is carried out by dividing the power consumption of the pump with the **Flow** (the amount of liquid) (see # 1 in the programming code).

Using a metaphor, the calculation can be explained by:

Power *is the price* you pay, and Flow *is what you get*.

```
FUNCTION_BLOCK DataCollect
VAR_INPUT
  Enabled:        BOOL:= FALSE; //  Start on positive trig
  Flow:           REAL; // Flow [m3/h]
  Power:          REAL; // Power [w]
END_VAR
VAR_OUTPUT
  ValueResult:  REAL; // Calculated result
  Done:           BOOL:= FALSE; // When TRUE ValueResult is ready
END_VAR
VAR
  EnableOneShot: R_TRIG;
  StateTimer:     TON; // Delay time before data is ready
END_VAR
```

```
//FUNCTION BLOCK for DataCollect
EnableOneShot (CLK:= Enabled);

//Start new data collect period. Can be restarted at any time
IF EnableOneShot.Q THEN
  ValueResult:= 1000; //Show high value if Flow <= 0
  Done:= FALSE;
  StateTimer (IN:= FALSE); //Reset timer
END_IF;

//Start delay timer, Period A..B
StateTimer (IN:= TRUE, PT:= T#10S);

//Timer end, collect value
IF StateTimer.Q THEN

  IF Flow > 0 THEN //Avoid division by zero
    ValueResult:= Power/Flow; //#1
  END_IF;

  Done:= TRUE;
END_IF;
```

```
PROGRAM MAIN
VAR
  DI_S1:          BOOL := FALSE;  // (8) Start DataCollect
  S1_OneShot:     R_TRIG;
  AO_MotorSpeed:  INT;            // (7)   Current speed
  MyDataCollect:  DataCollect;   //FUNCTION BLOCK
  AI_Flow:        REAL;           // (1)  Flow measure
  AI_Power:       REAL;           // (6)  Power measure
  DO_MotorStart:  BOOL;           // (5)  on/off for the motor
  LowValueFound:  REAL;
  LowSpeedFound:  REAL;
END_VAR
VAR CONSTANT
  SPEED_START:    INT := 85;   // Motor run speed at start
  SPEED_END:      INT := 100;  // Motor run speed at end
END_VAR

S1_OneShot (CLK:= DI_S1);

IF S1_OneShot.Q THEN      //Begin
  AO_MotorSpeed:= SPEED_START;
  DO_MotorStart:= TRUE; //Start Motor
  LowValueFound:= 10000;  //Init high value
END_IF;

//Loop all values of Speed
IF AO_MotorSpeed >= SPEED_START AND  AO_MotorSpeed < SPEED_END THEN
  MyDataCollect (Enabled:= TRUE, Flow:= AI_Flow, Power:= AI_Power);

  //Data collect done
  IF MyDataCollect.Done THEN
    AO_MotorSpeed:= AO_MotorSpeed + 1; //Set to next speed value
    MyDataCollect (Enabled:= FALSE); //End data collection

    //Save found value, if it is a lower than the old one #2)
    IF LowValueFound > MyDataCollect.ValueResult THEN
      LowValueFound:= MyDataCollect.ValueResult;
      LowSpeedFound:= AO_MotorSpeed; //Save current speed
    END_IF;
  END_IF;
ELSE //End of loop
  AO_MotorSpeed:= LowSpeedFound; //Set optimal speed
END_IF;
```

The function block uses the **Enable** input variable to start the measurement period. The **EnableOneShot** variable ensures that initialization of the variables is only performed when a new period starts.

In the **MAIN** program **DI_S1** starts a new series of measurements. All measurements are done without using the **FOR-DO** statement, because the function block has a timer that must be called at each program scan. The **FOR-DO** statement does not provide this kind of feature.

The loop starts at the minimum speed defined by the constant **SPEED_START**.The loop counter variable, **AO_MotorSpeed** is used and the measurements loop stops when **AO_MotorSpeed** is at the same speed as **SPEED_END**.

After each measurement period the new value is compared (see # 2 in the code) with the previous value. If the new value is lower than the previous value, the new value is saved in the **LowValueFound** variable. At the same time the speed is saved in **LowSpeedFound**.

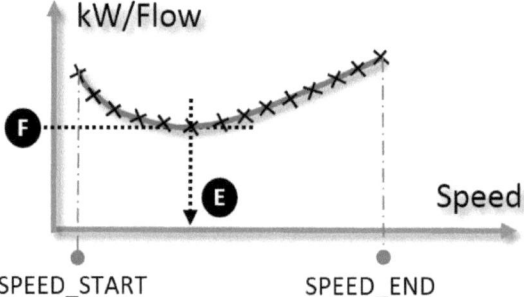

In the diagram, the lowest value is found at point (F). At this point the lowest cost per flow rate is found which is therefore the most optimal operating point for the pump.

The speed at the point (E) is the value in the variable **LowSpeedFound**, and this is the value to set the frequency converter at.

13.17 PLC control of Robot and CNC machine

This example shows a solution where items are processed in a CNC machine.

The plant consists of a table with 25 items, a robot, a CNC machine, a light tower and a start switch:

M1_P1 M2_P1
M1_P2 M2_Done S1
M1_Done

Control Description:

When the operator has pressed **S1**, the plant is in operation mode. This means:

> 1) The robot moves an item from the table to the CNC machine.
> 2) The CNC machine processes the item.
> 3) The robot moves the item back to the table.
> 4) When point 1 to 4 have been repeated 25 times, the job is done.

System design

It is important to decide whether it is the program in the robot controller or PLC that is responsible for ensuring that all 25 items are processed in the CNC machine. If the responsibility is placed with the robot controller, the robot controller must control the CNC machine directly, because the CNC machine has to notify the robot controller when it is ready for a new item. If the responsibility is placed with the PLC, the PLC controls both the robot and the CNC machine, which in many cases is a simpler design compared to a system with shared responsibility between the CNC and PLC.

To control the process, an **ARRAY** with 25 elements is declared:

```
VAR
   arTable: ARRAY[1..TableX, 1..TableY] OF BOOL;
END_VAR
```

```
VAR CONSTANT
   TableX: INT := 5;
   TableY: INT := 5;
END_VAR
```

The first step in the program sets all elements in the **ARRAY** to **FALSE**. When an item is moved to the CNC machine, the element is set to **TRUE**, to ensure the item is not selected again.

The robot gets a message from the PLC containing the position of the item to pick up. The message is sent via Fieldbus. The position is the contents of the variables **Take_x** and **Take_y**, multiplied by 120 [mm], because the robot uses the distance from the Home position to where the item has to be picked up.

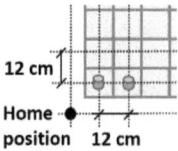

Because it is the PLC that decides which item the robot has to pick up, the PLC must also ensure that all 25 items are processed by the CNC machine.

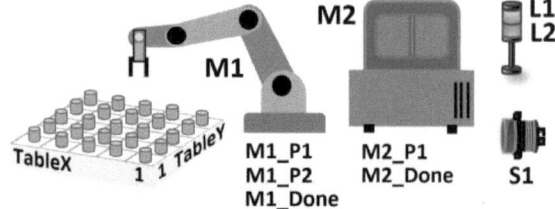

Components and how they work:

Name	I/O	Component	How do they work
S1	DI	Switch	The plant is in operation mode when pressing the switch
M1		Robot	Move items to and from CNC machine.
M1_P1	DO	Robot	**TRUE** when the robot starts robot program 1, where an item is moved from the table to the CNC machine.
M1_P2	DO	Robot	**TRUE** when the robot starts robot program 2, where an item is moved from the CNC machine back to the table.
M1_Done	DI	Robot	**TRUE** when the robot program is completed.
M2		Machine	CNC machine to process items (drill holes in the item).
M1_P2	DO	Machine	At signal (**TRUE**), CNC machine starts program 1. The machine takes the item supplied by the robot and processes it. Finally, signal is given on **M2_done**
M2_Done	DI	Machine	**TRUE** when the item can be removed from the CNC machine.
TableX		Table	X position on table with items.
TableY		Table	Y position on table with items.
L1	DO	Blue light	Light turned on when the plant is in operation mode.
L2	DO	Green light	Light turned on when the plant is ready for 25 new items.

The program is split into one main program and three **ACTIONs**:

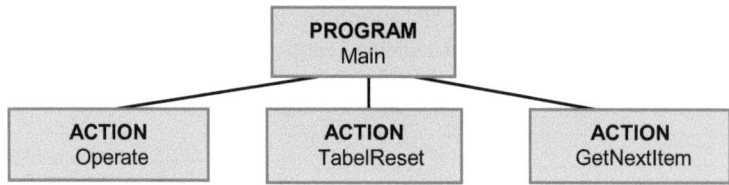

TableReset resets the **ARRAY** when starting a new table. **GetNextItem** gets the position of the next item from the **ARRAY**. All variables are created in the Main program and variables are shared with all three ACTIONs.

To obtain better program structure, **TableReset** and **GetNextItem** can be created as a function. If **GetNextItem** is a function, the two variables **Take_x** and **Take_y** must be grouped in a **STRUCT** so that the function can return the variables as a group.

If sequence programming (EN 60848) is used, **Main** and **Operate** will provide a better program structure, as the machine only processes one item at a time.

Main (PRG)

```
//Manual push button activated
IF S1 AND L2 THEN
  TableReset(); //Table full with items
  L2:= FALSE; //Turn off the task done lamp
  L1:= TRUE; //Turn on the operating lamp
  RunAll:= TRUE;    //Set to run for all items
  RunOne := FALSE; //Set to run for one item
END_IF;

//Take next item from the table
IF RunAll AND NOT RunOne THEN
  GetNextItem();
END_IF;

//If there are no more items, then stop
IF (Take_x = 0 AND Take_y = 0) THEN
  RunAll:= FALSE;
  L2:= TRUE; //Signal operation completed
  L1:= FALSE; //Turn off operating light
END_IF;

//One item is in operating mode
IF RunAll THEN
  Operate();
END_IF;
```

ACTION: Operate

```
//NOTE: Send Take_x and Take_y position to the
// robot before executing the code below

//Signal to Robot: Move one item
M1_P1:= TRUE;    //Start robot program 1
RunOne:= TRUE; //One item in progress

//Robot task done, start CNC
IF M1_Done AND M1_P1 THEN
  M1_P1:= FALSE;
  M2_P1:= TRUE;
END_IF;

//CNC task done, move item back to table
IF M2_Done THEN
  M2_P1:= FALSE;
  M1_P2:= TRUE; //Start robot program 2
END_IF;

//Robot task done,
IF M1_Done AND M1_P2 THEN
  M1_P2:= FALSE;
  RunOne:= FALSE;  //Stop item in progress
END_IF;
```

ACTION: TableReset

```
//New table
//Reset all elements in the ARRAY
//Set to FALSE indicates that the item
//is ready to be moved
FOR X := 1 TO TableX DO
  FOR Y := 1 TO TableY DO
    arTable[X, Y]:= FALSE;
  END_FOR;
END_FOR;
```

ACTION: GetNextItem

```
//Find the next item to be moved from the table

//Reset next item position to zero
Take_x:= 0;  Take_y:= 0;
Found := FALSE; //None found

//Loop all items
//Get the last item
FOR X := 1 TO TableX DO //All x positions
  FOR Y := 1 TO TableY DO  //All y positions
    IF NOT arTable[X, Y] AND NOT Found THEN
      Take_x:= X;  //Save x position
      Take_y:= Y;  //Save y position
      arTable[X, Y] := TRUE; //set as item to pickup
      Found:= TRUE; //A item is taken
    END_IF;
  END_FOR; //y loop
END_FOR; //x loop
```

14 From Ladder Diagram to ST-programming

This chapter contains a range of examples, comparing Ladder Diagram (LD) programming code with the corresponding ST programming code.

This chapter is meant to support the readers who understand LD programming well, or need to translate a LD program to ST. There are currently no tools available which are able to convert a LD program into an ST program, which is why the following examples have become part of this book:

Example 1: Input (contact) and output signal (coil)

```
       S1                                    K1
   ||  ||                                  ( )
```

```
//Solution 1A
K1:= S1;

//Solution 1B
IF S1 = TRUE THEN
  K1:= TRUE;
ELSE
  K1:= FALSE;
END_IF;
```

Example 2: Invert the inputsignal (Negation)

```
       S2                                    K2
   ||/||                                   ( )
```

```
//Solution 2A
K2:= NOT S2;

//Solution 2B
IF S2 = FALSE THEN
  K2:= TRUE;
ELSE
  K2:= FALSE;
END_IF;
```

Example 3: Input signal with one shot

```
       S1                                    K1
   ||P||                                   ( )
```

```
//Solution 3
VAR
  S1_TRIG: R_TRIG;
END_VAR
```

Example 3A: Input signal with **R_TRIG**

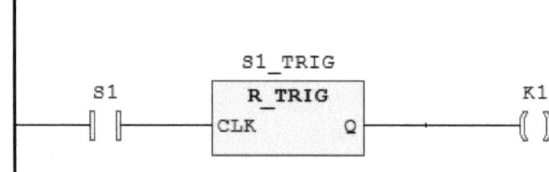

```
S1_TRIG (CLK:= S1);
IF S1_TRIG.Q = TRUE THEN
  K1:= TRUE;
ELSE
  K1:= FALSE;
END_IF;
```

Example 4: Latching relay / self hold relay

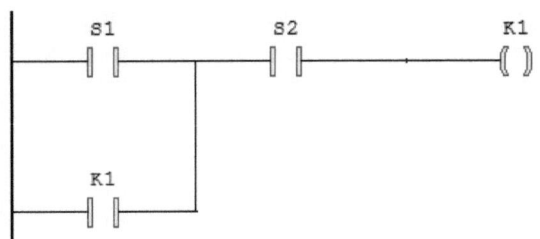

```
//Solution 4A
K1:= (K1 OR S2) AND S3;

//Solution 4B
IF ((K1 OR S2) AND S3) THEN
  K1:= TRUE;
ELSE
  K1:= FALSE;
END_IF;
```

Example 5: SET and RESET

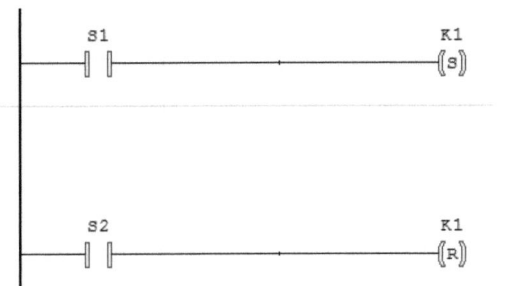

```
//Solution 5
IF S1 = TRUE THEN
  K1:= TRUE;
END_IF;

IF S2 = TRUE THEN
  K1:= FALSE;
END_IF;
```

Example 6: Timer

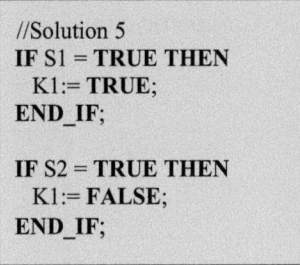

```
//Solution 6
MyTimer (IN:= S1, PT:= T#3S);
K1:= MyTimer.Q;
```

Example 7: Timer with automatic restart

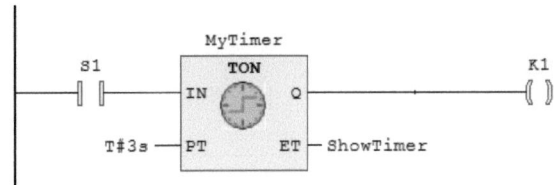

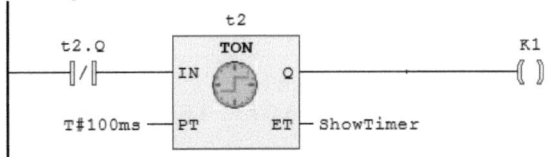

```
//Solution 7
t2 (IN:= NOT t2.Q, PT:= T#100ms);
K1:= t2.Q;
```

Example 8: More inputs and outputs

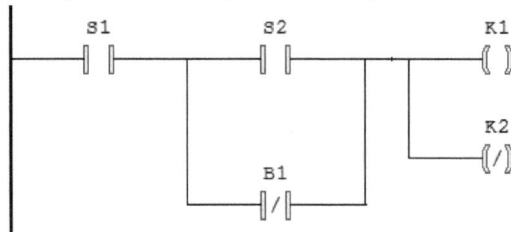

```
//Solution 8
IF (S1 AND (S2 OR NOT B1)) THEN
  K1:= TRUE;
  K2:= FALSE;
ELSE
  K1:= FALSE;
  K2:= TRUE;
END_IF;
```

Example 9: Compare values

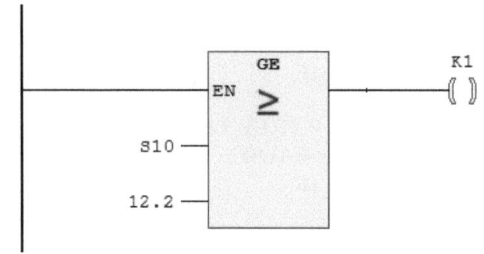

```
//Solution 9A
K1:= S10 >= 12.2;

//Solution 9B
IF S10 >= 12.2 THEN
  K1:= TRUE;
ELSE
  K1:= FALSE;
END_IF;
```

Example 10: Variable assignment (Move value to a variable)

```
//Solution 10A
IF (S1 = TRUE AND S2 = FALSE) THEN
  K2:= 123;
 END_IF;
```

```
//Solution 10B
IF S1 AND NOT S2 THEN
  K2:= 123;
 END_IF;
```

Example 11: Counter using the **CTU** function block

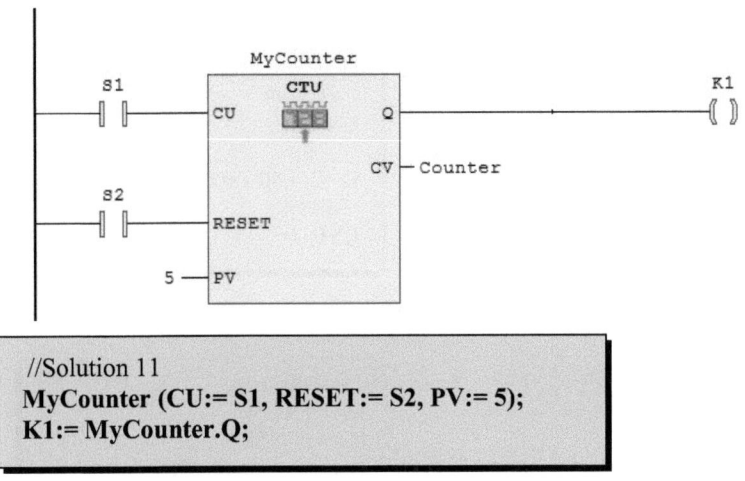

```
//Solution 11
MyCounter (CU:= S1, RESET:= S2, PV:= 5);
K1:= MyCounter.Q;
```

Example 12: Calculations

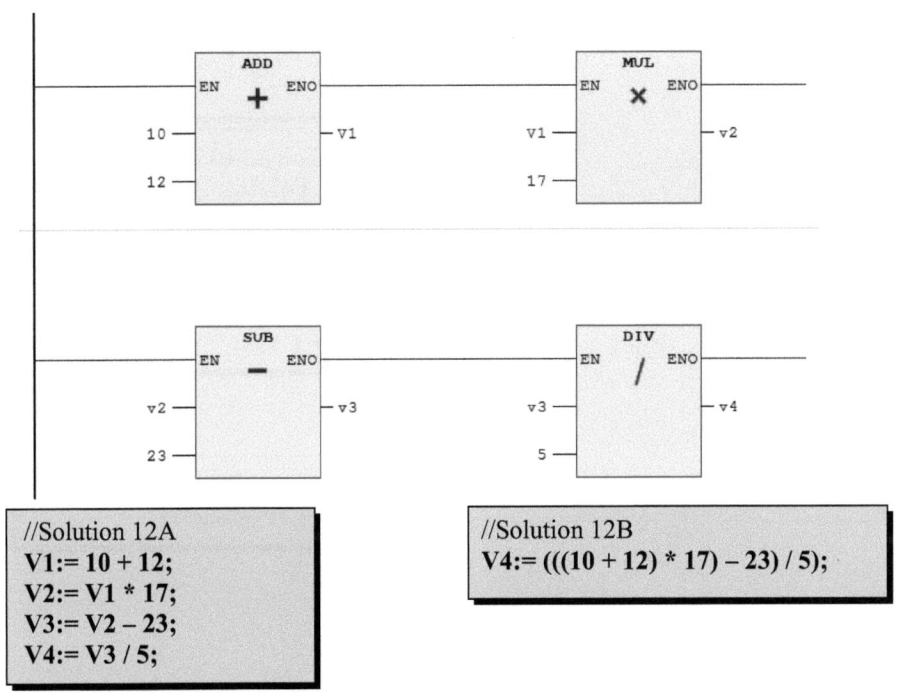

```
//Solution 12A
V1:= 10 + 12;
V2:= V1 * 17;
V3:= V2 – 23;
V4:= V3 / 5;
```

```
//Solution 12B
V4:= (((10 + 12) * 17) – 23) / 5);
```

Example 13: LIMIT function (value inside range 10 to 40)

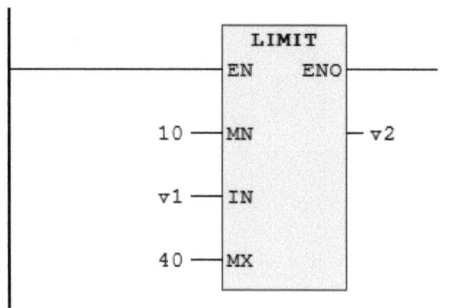

```
//Solution 13
v2:= v1;
IF v2 < 10 THEN
    v2:= 10;
END_IF;

IF v2 > 40 THEN
    v2:= 40;
END_IF;
```

Example 14: Compare values

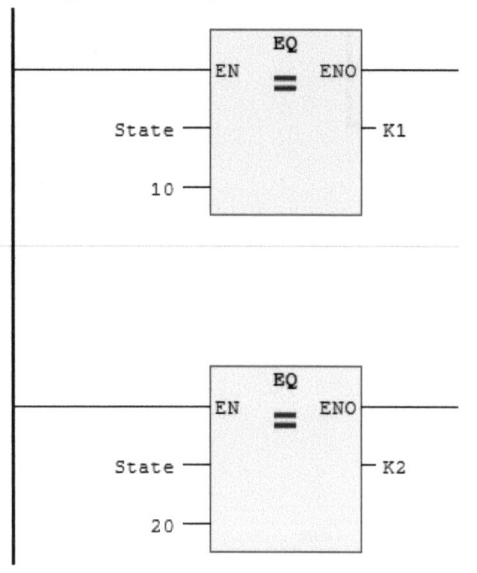

```
//Solution 14A
K1:= FALSE;
K2:= FALSE;
IF State = 10 THEN

    K1:= TRUE;
END_IF;

IF State = 20 THEN
    K2:= TRUE;
END_IF;
```

```
//Solution 14B
K1:= FALSE;
K2:= FALSE;

CASE State OF
    10: K1:= TRUE;
    20: K2:= TRUE;
END_CASE;
```

Example 15: Counter

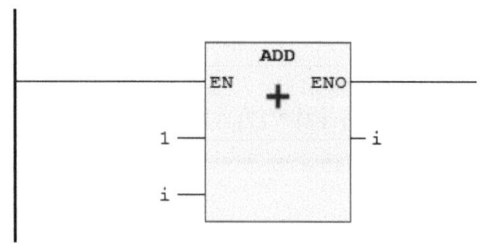

```
//Solution 15
i:= i + 1;  //Counter
```

Example 16: Trafic light

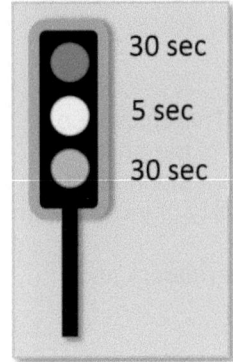

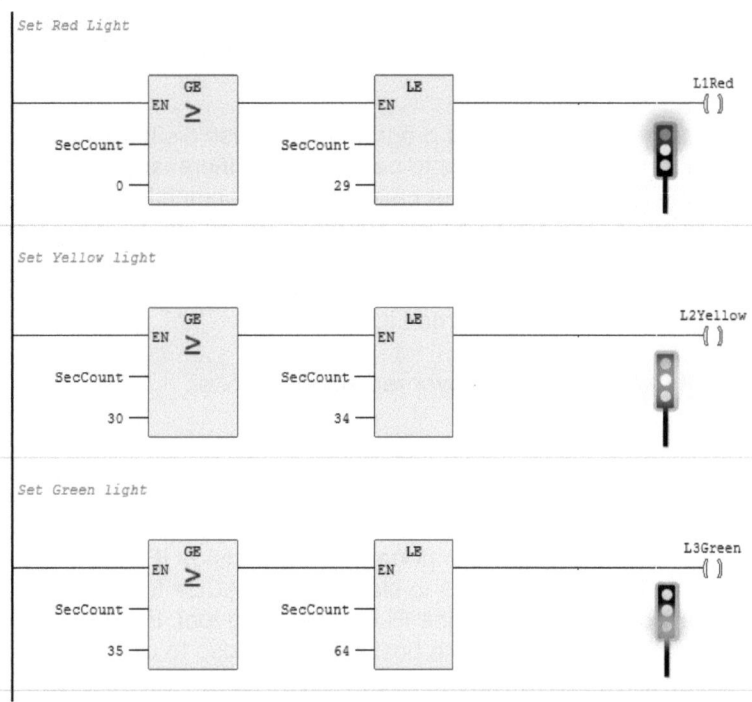

//Solution 16A
L1Red:=   FALSE;
L2Yellow:= FALSE;
L3Green:=   FALSE;

//Using case
CASE SecCount **OF**
  0 .. 29 : L1Red:=   **TRUE**;
  30 .. 34 : L2Yellow:= **TRUE**;
  35 .. 64 : L3Green:= **TRUE**;
END_CASE;

//Solution 16B
L1Red:= (SecCount >= 0) AND (SecCount <= 29);
L2Yellow:= (SecCount >= 30) AND (SecCount <= 34);
L3Green:= (SecCount >= 35) AND (SecCount <= 64);

//Solution 16C
L1Red:=   FALSE;
L2Yellow:= FALSE;
L3Green:= FALSE;

IF (SecCount >= 0) **AND** (SecCount <= 29) **THEN**
   L1Red:= **TRUE**;
END_IF;

IF (SecCount >= 30) **AND** (SecCount <= 34) **THEN**
   L1Yellow:= **TRUE**;
END_IF;

IF (SecCount >= 35) **AND** (SecCount <= 64) **THEN**
   L1Green:= **TRUE**;
END_IF;

15   Best Practice ST-programming

ST-programming enables programmers to use his/her own syntax. Therefore, good programming practice has to be followed to increase the readability of the entire program. Capital and lower case letters in conjunction with tabulation and indentation (placing of <SPACE>) in the program code can improve the readability of the program.

It is important to be consistent with your programming, so that other readers and programmers can easily read it.

The following sections cover recommendations:

15.1   Tabulation of text and placing of SPACE

Tabulation of the text is important for code inside **IF** and **CASE** statements and **FOR** loops. The best solution is to place 2 x <SPACE> for tabulation, because <TAB> depends on the setup in the PLC developing tool. If the PLC code must later be copied to another PLC, the best solution is also to use 2 x <SPACE>.

Tabulation increases the readability of the PLC code. The PLC code can be difficult to read without tabulation of the text or by using wrong tabulation. Therefore, be consistent with your tabulation throughout the entire program.

<SPACE> has no function in ST programming (the PLC ignores white space). To improve readability of the code place one <SPACE> between commands, variables, statements, brackets and values. It is, however, recommended not to place <SPACE> before semicolon.

15.2   Empty lines between code

It makes sense to create empty lines in the PLC code to separate and split up the different code pieces into smaller sections.

It is recommended to have a maximum of two empty lines between code sections.

Furthermore, do not place code lines on the same line as an ELSE-statement, and do not write code lines longer than the screen width.

15.3 Avoid spaghetti code

The spaghetti code is a definition for PLC code possessing a complex structure which occurs with unclear naming of variables and functions, many **GOTOs**, **JMPs**, **EXITs** or other unstructured implementations.

It is only recommended to use **GOTO** and **JMP** statements in very special situations (e.g. for fault finding, testing and debugging). The use of **EXIT** can also cause spaghetti code. It is, therefore, recommended to avoid the **EXIT** command and use other conditional statements such as **IF** and **CASE**. On the other hand, **EXIT** can be useful, when fault finding is carried out in the PLC code. Care must, however, be taken when the PLC code is finalized, and remember to remove the unused **EXITs**.

It is useful to use **EXIT** in FOR-loops, if it is not necessary to execute the whole loop. For implementation of **EXIT** in a **FOR**-loop see chapter 9.3, page 70.

GOTO and **JMP** are only available in some PLC types and if used the code cannot always be copied to another PLC.

15.4 Good program structure

The most basic way of implementing a good program structure is to split up the program in program modules and functions. By splitting up a large program into small programs, each with a specific task, it is possible to create a small **Main** program, 'calling' a number of program modules and functions.

Functions and function blocks are very effective; they can easily be reused in other programs and tested separately. Furthermore, when correcting the code, only one place has to be fixed. Usually, reusing functions and function blocks is easier than reusing program modules.

It is also recommended to give the functions and program modules an indicative name like variables, to ensure they are easy to recognize.

If a function or a program module contains more than 20 local variables, it is an indication that the program code, must be split up into several functions or program modules, or the numbers of variables should be reduced by using a **STRUCT**.

If possible, split the program into sequences according to the EN 60848 standard.

More than 40 lines of code in a program module or a function is not good structure.

A large program module can advantageously be split-up into ACTION modules.

15.5 The use of variables

When programming, a decision between using local and global variables must often be made. Using global variables is fast and easy to work with, because they are only declared once in a common variable/TAGS list. However, their use creates a bad program structure because all functions and program modules have access to the variables.

It is recommended to use local variables where possible, and delete the variables which are no longer in use. Limit the number of variables, by creating fewer variables with indicative and useful name.

Use **STRUCT** to group different variables in an object.

To prevent using up memory unnecessarily, only create **ARRAY**s with the length (size) needed.

If a function or a program module contains more than 20 local variables, it could be an indication of a bad program structure, and the program should be split up into functions.

15.6 Miscellaneous

Below are a few other programming tips and recommendations:

- Exchange complicated **IF-THEN** statements with a **CASE** statement
- Avoid **ELSIF** statements
- Avoid creating infinite loops. Consequently **DO-WHILE** are not recommended
- Do not use more than 3 incorporated loops in **FOR-DO** loops
- Each function must contain a max. of 20-25 lines of code – as much as you can see on the screen when programming and on a paper print out
- Do not use more than three-dimensional arrays (3D **ARRAY**)
- Use **CONSTANT** if the same number is used more than once
- Program modules or functions must contain a max. of 20 local variables

Avoid creating unnecessary **ARRAY** elements. They are easy to create, and unfortunately, some programmers create too many and too long elements, which uses system resources unnecessarily and creates a heavy load on the PLC

It is recommended to use parentheses in math formulas and calculations in algorithms to make sure that the calculation is correct.

15.7 Code sharing on the internet

Google offers unique value for programmers to find code on the internet. However, sometimes it takes a long time to find useful code, and it might include faults. As a result, it may be easier for you to write your code yourself. Copyright on the code can also be a reason why you cannot use it, if you or your company will make a profit from using the code.

Another challenge when finding code on the internet is that variables and structures often do not follow best practice for naming or programming. Often more time is spent on finding, correcting and adjusting code, compared to how long it would have taken to write the code yourself.

If you are employed in a company, be careful about uploading your own code to the internet. The code is the company's property and sharing it can be considered theft. Also make sure you know your company's policy concerning program code and participation with comments and in discussions of other companies' PLC codes and programming solutions in internet forums. Contributing in this way can go against 'The Employers' and Salaried Employees' Act' in Denmark and elsewhere, because your contribution might benefit others and not the company you work for.

15.8 OOP – Object-Oriented Programming

To structure the PLC code better, the philosophy from Object-Oriented-Programming (OOP) can be used. This means that variables and PLC code, which are related to each other, are collected in an object. Variables which are e.g. used for a motor are collected in a **STRUCT** (see chapter 4.4, page 20) and the operational conditions for a motor are collected in an **ENUM** (see chapter4.3, page 18).

Variables and constants which e.g. work on the same **ARRAY** can have the same first name to mark their relationship.

Functions and function blocks are created so they work in objects, this could be a sensor or an instrument, and the code can easily be reused elsewhere in the program. Some PLC types offer OOP, as described in the standard IEC 61131-3. These PLC types contain **METHOD** (mode of operation as a function), **ACTION** (mode of operation as a program module), **PROPERTY** and **TRANSITION**.

16 Guide and help during ST-programming

This chapter provides a guide that can be a help when writing a ST program. It also includes a guide for debugging and testing program code.

16.1 Guide to programming exercises

This chapter is a guide to help the reader when solving programming exercises.

1) Get started

Read the task and, as a rule of thumb, read it more than once. It is important only to solve exactly what the task describes and nothing more than that. The program specification has often been developed by the customer who is unlikely to pay extra for additional unrequested code. In addition, with more code comes more scope for error, which is often experienced as bad code quality by the customer.

If the task is not well described, it is important to examine any uncertainties within the specifications. Specifications are described in a document creating a total overview of how the whole automation solution has to work. The document is called a function description or control description. A good document will help retain knowledge and support communication with the customer to confirm the program specifications.

2) I/O-liste

Work out an I/O list. Study what the individual sensors and instruments have to measure and how they work. The I/O list is an important tool both during programming, testing, commissioning and the continuous maintenance and potential expansion of equipment at a later date.

It is important that the I/O list is more than 95% percent complete before starting the programming process, because changes in the I/O list can influence the programming and the subsequent tests.

Make sure that names for variables/TAGS already in the I/O list are indicative, as the names are used across the whole project, and the I/O list is a part of the documentation. If the task, diagram or document already contains indicative names, use these to make sure that they are identifiable.

3) HMI

Many PLC solutions include a user interface, which consists of an HMI (Human Machine Interface), electrical on/off switches and control lamps. Make a rough outline on paper/hard copy of how the picture layout could look like. Show the picture layout to the customer, the end user or a colleague to gain feedback. It is very time consuming to correct diagrams and illustrations later. Therefore, it is important that these are as correct as possible, before starting the configuration of the HMI.

You can also work out a list of which variables/TAGS will be exchanged between the HMI and the PLC program, because an interface description always provides a good overview. It might not be the same person who are configuration the HMI and programming the PLC. This is why a list is useful as guidance for both.

4) Flowcharts

Work out flowcharts for the complex program parts, so you have a better feeling of how the program must work. Flowcharts are good guidance for you and others who need to understand the program flow and how it works.

5) Design fase

Before starting the programming process, work out a design draft on paper, which contains the different program modules, functions and function blocks. This could take the shape of flowcharts which describes the program and can be part of the program design phase. This description also includes the names of the program modules and functions, briefly describing each program module and function. A certain level of experience is needed to be able to design a complete program before starting the programming phase, and therefore you might benefit from using the bottom-up method described in the next part of this chapter.

6) Programming and implementation

There are 2 ways to start the programming and implementation phase. The top-down and the bottom-up method.

The bottom-up method is a modular approach, where you start by writing small pieces of PLC code/functionality, which you know are required as parts of the program. You can start, by writing the code that you know is needed and is clear to you. If e.g. the program has a lamp which has to flash, write the PLC code which can make a lamp flash. Gradually a number of small PLC code lines are written – small building blocks. Through this process a lot of knowledge is gained, and step by step you get the feeling of how the entire program should function, and the individual building blocks work together.

The HMI can be used throughout the process to test small chunks of code, and to make sure the building blocks work together as components of a complete program. Keep testing the individual building blocks throughout the programming process, as it is more difficult to debug a large program. Tests of small programs are often called module tests, and their tests can be documented via e.g. screen dumps, so you can document to yourself and others that the program works well.

It might be helpful to work on two projects (in the PLC developing tool) at the same time. One project becomes the final solution, and the other project is for testing the individual program parts (a sandbox test). Small solutions in a project are tested, and when it works, it can be copied (copy pasted) or rewritten into the final project.

Run-time error can cause the PLC developing tool, executed in a Windows environment, to crash (the blue screen of death). So it is a good idea to save the PLC code often. This must also be done every time the PLC code works well, so you do not loose it.

If you are confused about how the individual program parts can be implemented, use Google for inspiration. However, sometimes more time is spent searching the internet rather than trying to code on your own. Remember, if you get stuck, do not use more than 15 minutes before you move on to another task. To use your time in the best possible way, ask your company's support department, a colleague or do a google search to get another pair of eyes on the problem. It is often little things that are missing from the code, and if you cannot solve it within 15 minutes, then you probably cannot solve it within 60 minutes.

16.2 Programming and troubleshooting tips

This chapter contains suggestions to help you troubleshoot your ST program in case of errors.

It can be challenging to troubleshoot (debug) in single steps and break points, because the PLC program runs in real-time and the program execution depends on variables based on sensor signals and timers. The PLC development tools provide features for single step troubleshooting and adding break points:

- Single step is when a program is executed one code line at a time.
- Break point is an intentional stop or pause in the program.

In most cases, it is easier to troubleshoot a program made up of functions and program modules, than one large program. The individual functions and program modules can be disabled during program execution by placing the signs // in front of the place where they are called from. This will reduce the program size and help locate errors.

Below find a list of tips and tricks that can help you when you program, troubleshoot and develop the PLC program:

1) Use two project files
Work with two projects files at the same time. Use one of the project files for the program you will eventually release. The other project file you can use to try out code, create demo programs and develop functions and function blocks.

2) Test your code frequently
Test your program on an ongoing basis. Write only 3 to 6 lines of code and then test the code immediately. If you write a lot of code without testing it, you might have to go through a lot of deleting, disabling code lines etc. to get to test the code you wrote first.

3) Working with timers (TON and TOF)
When working with timers, change the time setting so that you do not need to wait for the timer to expire. Remember to change the time back to the correct time setting afterwards.
It can help to copy the **ET** to a new variable to ensure that the timer is working well.

4) Working with ARRAY
The sizes of an ARRAY can be changed during testing and programming. It is easier to test code when the ARRAY has 5 elements compared to 1000 elements. Remember to set the ARRAY to the correct size before releasing your program.

5) Check if individual code parts are being executed

Insert this line "i: = i + 1;" to see if the program code has been executed. If the variable **i** counts 1 up, the program code is working well. This can be helpful to verify whether a program module or a function actually gets executed when the program is running.

The method can also be used to check execution inside an IF statement:

```
IF S1 THEN
  K1:= TRUE;
  i:= i + 1; //Is the code executed, i.e. does S1 get activated?
END_IF;
```

6) Use the HMI for testing and troubleshooting

Design a page for the HMI to be used only for testing and troubleshooting. Access to this page should be password protected to ensure that only you and the service technicians have access to the page. On the page show internal variables, sequence numbers, test code or program errors numbers etc.

7) Using test mode flag (test bit)

A test bit (**BOOL**) can be created and set to **TRUE** during the test and simulation of the program. It can be used to test a stop switch, which is an NC switch and therefore a **N**ormally **C**losed physical switch:

```
IF TestModeBit THEN
  S2_STOP:= TRUE;
END_IF;
```

Remember to set **TestModeBit** to **FALSE** after testing.

8) Programing code with logic: B1 AND B2 OR B3

Writing code like this can be a challenge:

```
K1:= (B1 AND B2) OR B3;
```

Using the syntax above can make troubleshooting difficult. Instead use IF statements:

```
K1:= FALSE;

IF (B1 AND B2) OR B3 THEN
  K1:= TRUE;
END_IF;
```

9) Troubleshooting complex program code

It can be helpful to insert an additional test variable when the program code is complex.

The variable can be used to check that the code inside an IF statement has been executed.

The contents of **TestVar** shows which IF statement that has been executed.

```
TestVar:= 1;  //Test

IF S1 AND S2 THEN
    TestVar:= 2; //Test
  IF B3 THEN
      TestVar:= 3; //Test
      K1:= TRUE;
  ELSE
      IF B4 THEN
          TestVar:= 4;  //Test
      K2:= TRUE;
      END_IF;
  END_IF;
END_IF;
```

10) Using log file for testing

In the previous point (9) a variable was used to check complex IF statements.

It can also be beneficial to use a **STRING** text as shown on the right:

The contents of **TestStr** can be shown on the HMI or written to a log file. As errors could occur at night, a log file is very useful because it can maintain a 24 hour log.

```
TestStr:= "Before IF";
IF S1 AND S2 THEN
    TestStr:= "Line 27 in MyDemo";
    K1:= TRUE;
END_IF;

IF B3 AND B4 THEN
    K1:= FALSE;
    TestStr:= "Line 30 in MyDemo";
END_IF;
```

11) If a variable does not change value

Try the following options:

Tip1: Insert a new variable and see if your code then works as expected. The variable has probably been changed elsewhere in the program code.

Tip2: Ensure there are "=" in the IF statements and not ":=". See chapter 9.1, page 56.

Remember if underline waves appear something needs to be fixed in the code!

16.3 Module test and simulation of connected equipment

It is important to test the functions, function blocks and program modules when they are developed, because it is difficult to find faults in a large program. It can be difficult to test functions and program modules as they often use signals from the connected components. The connected components can be sensors, instruments, conveyor belts and robots, which are often not available during testing.

Therefore, to perform a module test, it can be of great help to write small pieces of PLC code which can simulate the connected components.

This chapter shows how components can be simulated and used in testing.

1. Module test of a tank volume calculation

This is a test of the function described in chapter 13.11, page 150

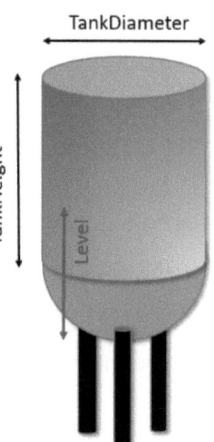

An array **ArTankTest** is created, where the calculated values can be stored. The tank diameter and height are created as variables, so it is easy to test different tank sizes.
It is important to test using different levels and levels outside the tank area value range to make sure the function works well.

```
PROGRAM MAIN
VAR
  TankD: REAL;  //TankHeight
  TankH: REAL;  //TankDiameter
  ArTankTest: ARRAY[0..7] OF REAL; //Calculated volume
END_VAR
```

```
//Set tank size
TankD:= 2;
TankH:= 4;

//Calculate volume at different levels
ArTankTest[0]:= TankVolumeCal(TankD, TankH, LevelFromBottom:= -1);
ArTankTest[1]:= TankVolumeCal(TankD, TankH, LevelFromBottom:= 0);
ArTankTest[2]:= TankVolumeCal(TankD, TankH, LevelFromBottom:= 1);
ArTankTest[3]:= TankVolumeCal(TankD, TankH, LevelFromBottom:= 1.5);
ArTankTest[4]:= TankVolumeCal(TankD, TankH, LevelFromBottom:= 2);
ArTankTest[5]:= TankVolumeCal(TankD, TankH, LevelFromBottom:= 3);
ArTankTest[6]:= TankVolumeCal(TankD, TankH, LevelFromBottom:= 4);
ArTankTest[7]:= TankVolumeCal(TankD, TankH, LevelFromBottom:= 5);
```

2. Simulation of box sizes

A plant contains a conveyor belt with different boxes:

The boxes can be simulated by creating an array that contains the different box sizes that may occur:

BoxArray

5	10	15	20	30	40	50

The boxes can come in random order on the conveyor belt. To simulate this, a random number generator is used (see chapter 13.3, page 132). The generator is set to give numbers that are in the range -3 to 3. To the random number the 3 is added so the random number range becomes 0 to 6. This number is index to **BoxArray.**

Below is the program code:

```
PROGRAM BoxSimulate
VAR
   BoxRND:     RND;      //Random number generator function block
   BoxIndex:   INT;      //Random generated index
   BoxTimer:   TON;      //Timer to create a time delay between each box
   BoxArray:   ARRAY [0..BOX_MAX ] OF INT := [5,10, 15, 20, 30, 40, 50]; //Box sizes
END_VAR
VAR_OUTPUT
   Box:        INT:= 0;  //Result Box
END_VAR
VAR CONSTANT
   BOX_MAX: INT:= 6;  //Equal sized array allowed only
END_VAR

BoxTimer(IN:= NOT BoxTimer.Q, PT:=T#10S); //Auto reset timer

IF BoxTimer.Q = TRUE THEN // Get a new box
   BoxRND(Seed:= 1, ValueMax:= BOX_MAX/2, ValueRandom => BoxIndex);
   BoxIndex:=  BoxIndex + BOX_MAX/2; //Always use a positive value
END_IF;

//Find the new box. BoxIndex have to be inside the array size
IF BoxIndex >= 0 AND BoxIndex <= BOX_MAX THEN
   Box:= BoxArray[BoxIndex];
END_IF;
```

Every 10 seconds, the **Box** variable contains a new size

3. Simulation of the liquid level in a pumping station well

It is often difficult to test a pumping station well with float switches, because the float switches are activated by different liquid levels.

This example shows how to simulate different levels:

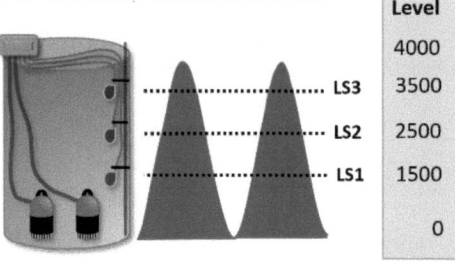

Level
4000
3500
2500
1500
0

The simulation of the levels is performed using a sinus curve (looks like a wave) simulation signal (see chapter 13.5, page 136). This signal oscillates (change) between minimum and maximum levels in the well.

In this example, the signal is a value between 0 and 4000. When the value reaches 1500, **LS1** is activated. When the value reaches 2500, **LS2** is activated, and at 3500 **LS3** is activated simulating the real world

Level simulation can be done by using this code:

```
PROGRAM LevelSimulate
VAR
   LS1, LS2, LS3 : BOOL    //Float switches in the well
   n:              REAL;   //To have a moving level in the well
END_VAR
VAR_OUTPUT
   Level:          REAL    //Level simulation
END_VAR
```

```
//Simulate level in the well
n:= n + 0.001; //Add for each program scan (depends on the task time)
Level:= 2000 + LREAL_TO_INT(2000 * SIN(n)); //Wave in the tank

//Active flow switches at different levels
LS1:= Level >= 1500;
LS2:= Level >= 2500;
LS3:= Level >= 3500;
```

The variable **n** counts 1 up for each program scan. Next, there are three conditional program lines which ensures that the float switches are activated at the right levels. **Level** is the simulated level in the well.

4. Simulation of a robot

If the PLC is used to control external equipment, PLC code can be written to simulate the equipment. In this example, the equipment is a robot controlled by a robot controller. The robot controller is a standalone computer that controls the robot's movements. The robot controller is controlled by digital signals:

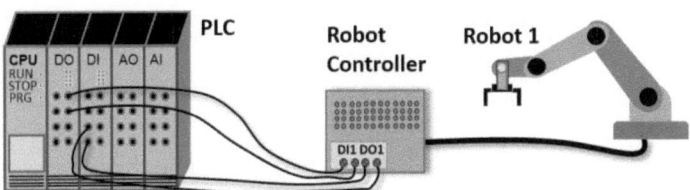

The robot controller has a digital input **DI1** which starts the robot program. When the robot program is finished, the robot controller sends a signal to **DO1**.

The robot controller can be simulated by using a function block, so that the PLC program can be tested without being connected to the robot controller.

In this example, the execution time of the robot program (cycle time) is 7 seconds:

```
FUNCTION_BLOCK RobotSim
VAR_INPUT
   DI1:        BOOL; //Start robot program
   PrgCycle : TIME; //Execution time
END_VAR
VAR_OUTPUT
   DO1: BOOL;        //Robot program done
END_VAR
VAR
   StartSim: BOOL := FALSE;
   TimeSim: TON;    //Robot work timer
END_VAR
```

```
//FUNCTION BLOCK RobotSim
IF DI1 THEN
   StartSim:= TRUE;
   DO1:= FALSE;
END_IF;

TimeSim(IN:= StartSim, PT:= PrgCycle);
//Robot program done

IF TimeSim.Q THEN
   StartSim:= FALSE;
   DO1:= TRUE;
END_IF;
```

```
PROGRAM MAIN
VAR
   Robot1: RobotSimu; //Function block for robot simulattion
   DI1:  BOOL; //Simulate digital input to robot controller
   DO1:  BOOL; //Simulate digital output from robot controller
END_VAR

Robot1(DI1:= DI1, PrgCycle:= T#17S, DO1=>DO1);
```

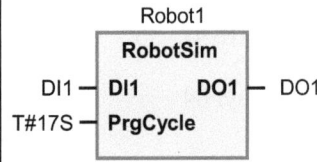

17 Index

B

C

D

E

F

FALSE; 44; 56
FBD; 6
Fieldbus; 11; 28
FIFO; 129
filtered signals; 135
FIND; 101
Finite Impulse Response; 134
firmware; 11
first order lag filter; 135
FirstScanBit; 108
Flashing Light; 119
Float; 15
FLOOR; 42; 50
Flowcharts; 57; 189
FOR loop; 128
FOR-DO; 70
formula; 45
Fourier transformation; 134
FRAC; 42
frequency converter; 158
FUNCTION; 83
Function (FC); 82
Function block (FB); 82
FUNCTION_BLOCK; 83
Functions; 80

G

Get a value from an array; 24
GOTOs; 185
GRAD; 43

H

hard coded; 49
Heartbeat; 47
heating process; 158
Hertz; 20
HEX; 14; 17; 51

high-level programming language; 6
HMI; 28; 37; 95; 146; 166; 189
Home position; 175
Hungarian Notation; 32

I

I/O list; 188
IEC 61131; 5; 14
IEC day; 15
IEC time; 15
IEEE Floating; 15
IF statement; 56
IF-THEN-ELSE; 56
Implementation of a function; 83
IN; 81
IN_OUT; 81
INC; 42
Increment; 42
indirect addressing; 25
input variable; 30
INSERT; 100
INT; 14; 49
INT_TO_BOOL; 52
INT_TO_REAL; 52
INT_TO_TIME; 52
Integer; 14
Invalid signs; 31
IO-Lists; 33
ISA-S88; 150
ISO 10646; 16
Iteration statement; 70
iterative variable; 34

J

JMPs; 185

K

keep relay; 60

L

Ladder Diagram; 6; 178
Language texts; 94
Latching relay; 179
LEFT; 101
LEN; 101
level measurement; 90
Line comments; 12
linear scaling; 92
liquid height; 147
LN; 43; 48
Local Temp; 28
local variables; 28
localized characters; 13
LOG; 48
Logic operators; 44
Loops; 70
Low Pass LP-Filter; 134
LOWER_BOUND; 25; 107
LREAL; 15
LTON; 117

M

manually operated; 62
mapping address; 29
Masking bit; 53
math functions; 42
mathematical rules; 45
mathematical symbols; 39
MES; 126
METHOD; 187
MID; 101
minimum value; 76
MIS; 126
MOD; 39
Module test; 194

Modulo; 39
MUL; 39
Multiply; 39

N

natural logarithm; 43
NEG; 42
Negation; 178
Noise Attenuation; 116
Normally **C**losed (NC); 60; 62; 142
Normally **O**pen (NO); 60
NOT; 44

O

Object-Oriented-Programming; 187
Off-delay timer; 109; 116
on/off signals; 120
ON/OFF switch; 56
On-delay timer; 116
one dimensional array; 22; 24
One Shot;109; 163
Online changes of languages; 94
OOP; 21; 150; 187
operational modes; 66
OR; 44; 45
OSF, OSR; 109
OUT; 81
output card; 30
output variable; 30

P

PAC; 5
parallel-connected components; 44
parentheses; 45
Pascal Case; 32
Pascal Programming; 6
password; 69
pause a timer; 125

T

T#; 116
TACHO HOURS; 17; 82
TAGS; 31; 33; 186; 188
TAN; 43
tank control; 158
temperature; 87
test bit; 192
TIME _OF_DAY; 15
timer delay; 116
TOF; 116
Toggle switch; 163
TON; 116
Top-down design; 85
Trafic light; 183
Triangle wave; 136
troubleshoot; 191
TRUE; 44; 56
TRUNC; 42; 50
two-dimensional; 23
two-way switch; 163

U

UINT; 14
unit; 36
UPPER_BOUND; 25; 107; 165
upper-case letters; 31

User defined data types; 18
UTC; 52

V

valve matrix; 54
VAR; 28
VAR CONSTANT; 130
VAR_GLOBAL; 28
VAR_IN_OUT; 28
VAR_INPUT; 28
VAR_OUTPUT; 28
VAR_TEMP; 28
variable; 46
variable creation; 30
variable names; 31
variable scope; 28
Variables with unit; 36

W

warehouse rack; 26
Watt Meter; 115
WCONCAT; 102
WCS; 126
WORD; 14
WSTRING; 95

X

X, Y and Z system; 23
XOR; 44